Autodesk Official Training Guide

Essentials

AutoCAD®

Civil 3D® 2010

Learning **AutoCAD**® Civil 3D® 2010

Using hands-on exercises, explore the essential elements for creating, analyzing, and managing civil engineering drawings and projects.

Autodesk®

Published by: Autodesk, Inc.
111 McInnis Parkway
San Rafael, CA 94903, USA

Special Thanks

Luke Pauw
Sr. Graphic Designer

Diane Erlich
Graphic/Production Designer

Daniel Gottlieb
Graphic/Production Designer

Elise O'Keefe
Copy Editor

Alan Harris
Technical Editor

Peter Verboom
Video Producer

Table of Contents

Chapter 01 | AutoCAD Civil 3D Environment

Lesson 01 | The User Interface   2

   Exercise 01 | Work in the User Interface   9
   Exercise 02 | Work with Workspaces   13

Lesson 02 | Examining Toolspace   16

   Exercise 01 | Examine Toolspace   21

Lesson 03 | Creating Objects, Object Styles, and Label Styles   28

   Exercise 01 | Create an Object Style   39
   Exercise 02 | Create a Label Style   43

Lesson 04 | Creating Drawing Templates   46

   Exercise 01 | Explore Command Settings   53
   Exercise 02 | Create a Drawing Template   57

Lesson 05 | Creating Reports   62

   Exercise 01 | Create Reports   67

Chapter 02 | Working with Survey

Lesson 06 | Creating Survey Databases   76

   Exercise 01 | Create a Survey Database   83

Lesson 07 | Creating Survey Networks   86

   Exercise 01 | Create a Survey Network   91

Lesson 08 | Creating Figure Styles and Prefixes   94

   Exercise 01 | Create Figure Styles and Prefixes   100

Lesson 09 | Importing Survey Data   104

   Exercise 01 | Review and Import Field Book Files   110

Lesson 10 | Working with Survey Data   114

   Exercise 01 | Work with Survey Data   120

Chapter 03 | Points

Lesson 11 | Importing and Creating Points   132

   Exercise 01 | Import Points from a Text File   142

Lesson 12 | Managing Points   146

   Exercise 01 | Create a Description Key   153
   Exercise 02 | Create Points Manually   155
   Exercise 03 | Create Point Groups   159

Chapter 04 | Surfaces

Lesson 13 | Creating Surfaces   166

   Exercise 01 | Create a Surface   172

Lesson 14 | Modifying Surfaces   176

   Exercise 01 | Modify Surface Properties   185
   Exercise 02 | Edit a Surface   189

Lesson 15 | Creating Surface Styles   192

   Exercise 01 | Create Surface Styles   196

Chapter 05 | Site Design Parcels

Lesson 16 | Creating Sites   202

    Exercise 01 | Create a Site   207

Lesson 17 | Creating Right-of-Way Parcels   210

    Exercise 01 | Create ROW Parcel   214

Lesson 18 | Creating Parcels   218

    Exercise 01 | Create Parcels Using Layout Tools   226
    Exercise 02 | Create Parcels from Objects   234

Lesson 19 | Editing Parcels   238

    Exercise 01 | Edit Parcels   242
    Exercise 02 | Renumber Parcels   246

Lesson 20 | Labeling Parcel Segments and Creating Tables   248

    Exercise 01 | Label Parcel Segments   253

Chapter 06 | Site Design Alignments

Lesson 21 | Creating Alignments from Objects   260

    Exercise 01 | Create Alignments from Objects   263

Lesson 22 | Labeling Alignments and Creating Tables   268

    Exercise 01 | Label Alignments and Create a Table   271

Chapter 07 | Site Design Profiles

Lesson 23 | Creating Surface Profiles and Profile Views   280

    Exercise 01 | Create a Surface Profile and a Profile View   285

Lesson 24 | Creating Layout Profiles   290

    Exercise 01 | Create a Layout Profile   295

Lesson 25 | Editing Profile Geometry   304

    Exercise 01 | Edit Profile Geometry   310

Lesson 26 | Labeling Profiles and Profile Views   314

    Exercise 01 | Label Profiles and Profile Views   319

Chapter 08 | Site Design - Assemblies and Corridors

Lesson 27 | Creating Assemblies   326

Exercise 01 | Create Assemblies   333

Lesson 28 | Creating Corridor Models   342

Exercise 01 | Create a Corridor Model   351

Lesson 29 | Creating Corridor Surfaces   360

Exercise 01 | Create a Corridor Surface   364

Chapter 09 | Site Design - Grading and Quantities

Lesson 30 | Creating Feature Lines   368

Exercise 01 | Create and Edit Feature Lines   371

Lesson 31 | Creating Interim Grading Surfaces   380

Exercise 01 | Create an Interim Grading Surface   385
Exercise 02 | Create a Grading Footprint   388

Lesson 32 | Creating Final Grading Surfaces and Calculating Volumes   396

Exercise 01 | Create Grading Objects   404
Exercise 02 | Create Final Grading Surface and Calculate Volumes   411

Lesson 33 | Labeling Final Grading Surface   414

Exercise 01 | Label Final Grading Surface   419

Chapter 10 | Site Design - Pipes

Lesson 34 | Creating Pipe Networks   427

  Exercise 01 | Create a Pipe Network   441

Lesson 35 | Drawing and Editing Pipe Networks   448

  Exercise 01 | Draw Pipes in Profile View   453
  Exercise 02 | Edit a Pipe Network   455

Lesson 36 | Labeling Pipes   460

  Exercise 01 | Label Pipes   463

Lesson 37 | Designing Storm Sewer Networks   468

  Exercise 01 | Design a Storm Sewer Network   475

Chapter 11 | Transportation - Alignments

Lesson 38 | Designing Criteria-Based Alignments   484

  Exercise 01 | Create and Edit an Alignment   492

Lesson 39 | Applying Superelevation   504

  Exercise 01 | Apply Superelevation   508

Lesson 40 | Creating Offset Alignments   512

  Exercise 01 | Create Offset Alignments and Widenings   518

Chapter 12 | Transportation - Assemblies and Corridors

Lesson 41 | Creating and Modifying Transportation Assemblies   528

  Exercise 01 | Create and Modify a Transportation Assembly   538

Lesson 42 | Creating Transportation Corridors   544

  Exercise 01 | Create a Corridor Model   550
  Exercise 02 | Map Corridor Targets   554
  Exercise 03 | View and Edit Corridor Sections   557

Lesson 43 | Creating Transportation Corridor Surfaces   562

  Exercise 01 | Create Corridor Surfaces   565

Lesson 44 | Creating Intersections — 570

Exercise 01 | Create an Intersection — 581

Lesson 45 | Modeling Road Designs in 3D — 588

Exercise 01 | Create a 3D Road Design Model — 592

Chapter 13 | Transportation - Sections and Quantities

Lesson 46 | Creating Sample Lines — 598

Exercise 01 | Create and Edit Sample Lines — 604
Exercise 02 | Modify Sample Line Group Properties — 608

Lesson 47 | Calculating Corridor Quantities — 610

Exercise 01 | Calculate Corridor Quantities — 615

Lesson 48 | Creating Quantity Reports — 620

Exercise 01 | Create Quantity Reports — 624

Lesson 49 | Creating Section Views — 628

Exercise 01 | Create Multiple Section Views — 632

Chapter 14 | Manage Data

Lesson 50 | Plan Production — 639

Exercise 01 | Create Construction Plans — 644

Lesson 51 | Working with Data Shortcuts and Reference Objects — 650

Exercise 01 | Work with Data Shortcuts and Reference Objects — 654

Lesson 52 | Calculating Quantity Takeoff Using Pay Items — 664

Exercise 01 | Calculate Quantity Takeoff Using Pay Items — 670

Lesson 53 | Working with Autodesk Vault — 678

Exercise 01 | Work with Vault — 690

Introduction

Welcome to the Learning AutoCAD Civil 3D 2010: Autodesk Official Training Guide, a training guide for use in Authorized Training Center (ATC®) locations, corporate training settings, and other classroom settings.

Although this guide is designed for use in instructor-led courses, you can also use it for self-pacedlearning. This guide encourages self-learning through the use of the AutoCAD® Civil 3D® Help system.

This introduction covers the following topics:

- Course Objectives
- Prerequisites
- Using this guide
- Installing the exercise data files
- Imperial and metric datasets
- Feedback

This guide is complementary to the software documentation. For detailed explanations of features and functionality, refer to the Help in the software.

Course Objectives

After completing this course, you will be able to:

- Describe the AutoCAD Civil 3D working environment.
- Use Survey functionality in Civil 3D.
- Create and manage points.
- Create and edit surfaces.
- Create, edit, and label, sites and parcels.
- Create and label alignments and create tables.
- Create surface profiles, profile views, and layout profiles. Edit profile geometry and label profiles and profile views.
- Create assemblies, corridor models, and corridor surfaces.

- Create feature lines, interim and final grading surfaces, and calculate quantities.

- Create, edit, and label pipe networks, and design storm sewer networks.

- Design criteria-based alignments, apply superelevation, and create offset alignments.

- Create transportation assemblies, corridors, and corridor surfaces. Create intersections and model road designs in 3D.

- Calculate corridor quantities, create quantity reports, and create section views.

- Produce plans, work with data shortcuts and reference objects, calculate quantity takeoff, and work with Autodesk Vault.

Prerequisites

This guide is designed for new AutoCAD Civil 3D users. It is recommended that you have:

- AutoCAD Civil 3D 2010 installed on your computer.

- Knowledge of civil engineering principles and processes.

- A working knowledge of AutoCAD®.

- A working knowledge of Microsoft® Windows® XP or Microsoft® Windows® Vista.

Using This Guide

The lessons are independent of each other, although they follow a typical civil engineering work flow. We recommend that you complete these lessons in the order that they are presented.

Each chapter contains:

- **Lessons**
  Usually two or more lessons in each chapter.

- **Exercises**
  Practical, real-world examples for you to practice using the functionality you have just learned. Each exercise contains step-by-step procedures and graphics to help you complete the exercise successfully.

Installing the Exercise Data Files

To complete the exercises in this guide, you must download the data files from the following location and install them on your system.

To install the data files for the exercises:

1   Download the zip file from *www.sybex.com/go/learningcivil3d2010*
2   Unzip the file Setup.exe.
3   Double-click Setup.exe and follow the onscreen instructions to install the files.
4   After the install is complete, you can delete Setup.exe from your system (optional).

Unless you specify a different folder, the exercise datasets are installed in the following folder: *C:\Autodesk Learning\AutoCAD Civil 3D 2010\Learning*

Download a Trial Version of AutoCAD® Civil 3D®

This guide was designed for use with AutoCAD® Civil 3D® 2010 software. If you do not have AutoCAD Civil 3D 2010 software installed on your system, you can download a trial version.

To download the latest trial version of the AutoCAD Civil 3D software:

1   Navigate to www.autodesk.com/autocadcivil3dtrial
2   Complete the registration and mailing information.
3   Submit the online form to download a free** 30-day trial version.

** This product is subject to the terms and conditions of the end-user license agreement that accompanies the software.

Imperial and Metric Datasets

In exercises that specify units of measurement, alternative files are provided as shown in the following example:

- Open *I_Pipe Networks.dwg* (imperial) or *M_Pipe Networks.dwg* (metric).

  In the exercise steps, the imperial value is followed by the metric value in parentheses as shown in the following example:

- For Length, enter **13'2"** (**4038** mm).

  In the exercise steps, the unitless value is specified as shown in the following example:

- For Length, enter **400**.

Feedback

We always welcome feedback on Autodesk Official Training Guides. After completing this guide, if you have suggestions for improvements please send your comments to:
learningtools@autodesk.com.

Chapter 01
AutoCAD Civil 3D Environment

This chapter describes the AutoCAD Civil 3D environment. To work effectively with Civil 3D®, you must understand the elements that make up the user interface, and how the fundamental methodology behind the program works.

The User Interface lesson introduces you to ribbons and workspaces. Ribbons organize commands in a process oriented way, and workspaces control the appearance and position of windows, menus, toolbars, and other user interface elements. The Examining Toolspace lesson describes how you use the different tabs to manage drawing data, drawing settings, and create reports. The Creating Object, Object Styles and Label Styles lesson describes objects and how you use styles to control object and label display. The Creating Drawing Templates lesson describes how you create a drawing template DWT file. Finally, the Creating Reports lesson describes how you create external reports for the data you have in your drawing.

▶ Objectives

After completing this chapter, you will be able to:

- Explain how you work with and customize the Civil 3D user interface.

- Describe how you use Toolspace to efficiently manage your drawings.

- Create object and label styles.

- Create drawing templates.

- Create reports using the Reports Manager.

Lesson 01 | The User Interface

This lesson describes the user interface in Civil 3D and explains how you manage the user interface to maximize your productivity.

Civil 3D is a complex design and drafting environment. Users work with many interface components to accomplish design and drafting tasks. When used properly, the final drafting and production of engineering and construction drawings is a by-product of the design process.

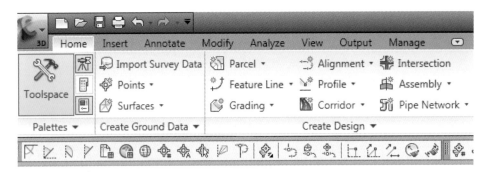

Civil 3D user interface

Objectives

After completing this lesson, you will be able to:

- Explain the purpose of static and contextual ribbons.
- Describe the AutoCAD workspaces in Civil 3D.
- Use the user interface to open files and display static and contextual ribbons.
- Explore existing workspaces and create a custom workspace.

About Ribbons

In AutoCAD Civil 3D, the ribbon is the primary user interface for accessing commands and features. While the traditional menus are still available, all commands for both AutoCAD® and Civil 3D are available on the ribbon.

Definition of Ribbon

The ribbon is a palette that displays task-based commands and controls. It is automatically displayed when you create or open a drawing file. The ribbon is comprised of tabs, panels, and commands. The tabs contain panels and the panels contain commands.

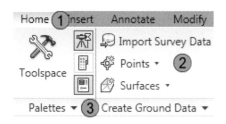

Ribbon components

① Tabs

② Commands

③ Panel

Ribbon Types

Ribbons are classified as either static or contextual ribbons.

Static Ribbon

Static ribbons are always displayed, and contain the tabs, panels and commands that you use most often. In Civil 3D, the static ribbon displays the Home, Insert, Annotate, Modify, Analyze, View, Output, and Manage tabs.

Contextual Ribbon

A contextual ribbon tab is displayed when you select an object in the drawing area or execute certain commands. It identifies the object, and shows panels and commands that can be used to work with the selected object. Contextual ribbons display only the applicable commands, making it easier to work with your data.

Examples

The following illustrations show examples of static and contextual ribbons.

Static Ribbon

Civil 3D uses the Home, Insert Annotate, Modify, Analyze, View, Output, and Manage tabs. When you select a tab, the ribbon displays the panels associated with that tab.

The Home tab contains the Palettes, Create Ground Data, Create Design, Profile & Section Views, Draw, Modify, Layers, Clipboard, and View panels. The Insert tab contains the Import, Block, Attributes, Reference, and Data and Linking & Extraction panels. The panels and their commands directly relate to the name of the tab.

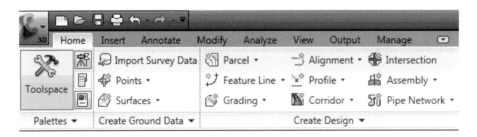

Contextual Ribbons

When a corridor model is selected in the drawing area, the contextual ribbon displays the name of the corridor, as well as the commands associated with corridors.

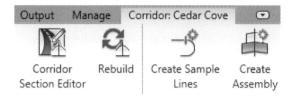

When a surface model is selected in the drawing area, the contextual ribbon displays the name of the surface, as well as the commands associated with surfaces.

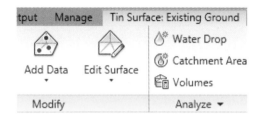

About Workspaces

Civil 3D has several predefined workspaces. You can use these workspaces as they are, or copy and modify them.

Definition of Workspaces

The workspace is the configuration of the user interface. Workspaces are sets of ribbons, menus, toolbars, and their positions, which are grouped and organized so that you can work in a custom, task- oriented drawing environment. When you use a workspace, only the ribbons, menus, toolbars, and secondary windows specified in that workspace are shown.

Note you can access commands not shown in the menus by entering their command names at the command line.

Modifying Workspaces

You can modify workspaces to add or remove toolbars, menus, and ribbon tabs and panels. You make these modifications in the Customize User Interface dialog box, as shown in the following illustration.

Civil 3D

This workspace displays the ribbon tabs, panels, and other interface components that show all Civil 3D related commands. You use the Civil 3D workspace to create site and transportation designs.

Workspace Examples

The following workspaces are included in Civil 3D:

Civil 3D

This workspace displays the ribbon tabs, panels, and other interface components that show all Civil 3D related commands. You use the Civil 3D workspace to create site and transportation designs.

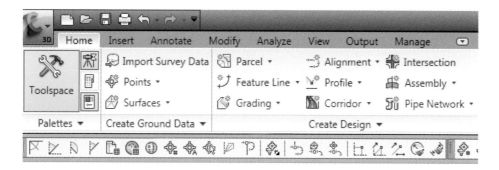

2D Drafting and Annotation

This workspace displays the ribbons tabs, panels, and other interface components required for two dimensional drafting and annotation tasks. You use this workspace to produce engineering and construction drawings.

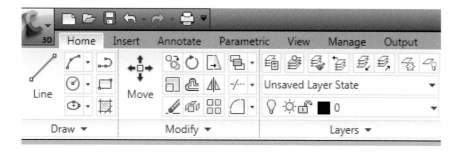

3D Modeling

This workspace displays the ribbon tabs, panels, and other interface components required for three dimensional modeling. You use this workspace to create 3D rendered models and animations that show your proposed design.

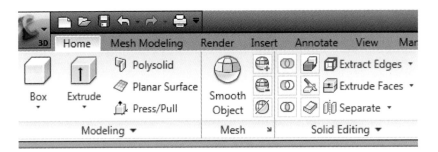

Workspace Switching

You can switch to a different workspace at any time by using the Workspaces Switching command. This is an AutoCAD command that is displayed on the application status bar menu at the bottom of the AutoCAD screen. The following illustration shows the Workspaces Switching command.

Exercise 01 | Work in the User Interface

In this exercise, you use the user interface to open files and display static and contextual ribbons.

The completed exercise

Open Files

1   Launch Civil 3D Imperial (Metric). Review the available options on the Welcome screen.

2   Close the Welcome screen.

Open files using the Application menu and the Quick Access toolbar.

3   To open a file using the Application menu:
    • Click the Application menu.

    • Click Open.
    • Browse to and open ...\AutoCAD Civil 3D Environment\user_interface.dwg.

4   To review the changes to the Application menu:
    • Click the Application menu again.
    • Review the available commands.
    • Click Close.

5   To open files using the Quick Access toolbar:
    • On the Quick Access toolbar, click Open.

    • Browse to and open ...\AutoCAD Civil 3D Environment\user_interface.dwg.

6   Review the changes to the Application menu.

Static Ribbons

Next, you display the commands within the static ribbons.

1   Click the Insert tab to display the panels and commands related to inserting.

2   Click the Modify tab.

3   On the panel, click Design to display the other commands on the Design panel.

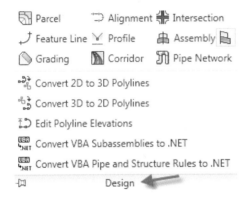

4   Click the Analyze tab.

5   Click the Volumes command to display the commands for volume calculation.

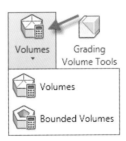

6   Click the Home tab.

7   Click the down arrow, to the right of the Manage tab, to turn off the panel commands.

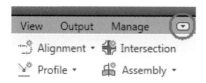

8   Click the down arrow again to turn off the panels.

9   Click the down arrow again to turn on the panels and the panel commands.

Contextual Ribbons

Finally, you examine contextual ribbons.

1   In the drawing area, click any contour to select the Existing Ground surface model.

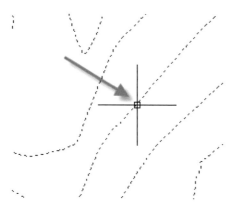

2   The contextual ribbon updates to display the name of the selected object.

The contextual ribbon also updates to display panels and commands applicable to surfaces.

3   Press ESC to deselect the surface. The static ribbon is displayed.

4   In the drawing area, click the cul-de-sac on the right to select the Cedar Cove corridor model.

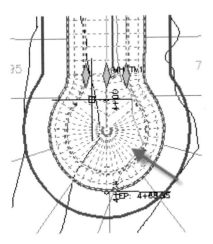

5   The contextual ribbon updates to display the name of the selected object.

6   The contextual ribbon also updates to display panels and commands applicable to corridor objects.

7   Press ESC to deselect the corridor object.

8   Repeat steps 4 through 7, selecting other objects in the drawing area.

9   Close the drawings. Do not save the changes.

Exercise 02 | Work with Workspaces

In this exercise, you explore the workspaces available in Civil 3D and modify some workspace settings. You then create a custom workspace by adding toolbars and removing menus.

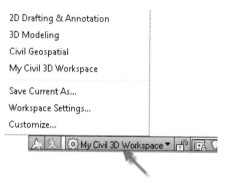

2D Drafting & Annotation
3D Modeling
Civil Geospatial
My Civil 3D Workspace

Save Current As...
Workspace Settings...
Customize...

The completed exercise

Review Workspaces

First, you review the existing workspaces.

1   Start Civil 3D Imperial (Metric).

2   Close the Welcome screen.

3   To switch to the 2D Drafting and Annotation workspace:
   • On the Application status bar menu, at the bottom right hand corner of the AutoCAD screen, click Workspace Switching.

   • Select 2D Drafting and Annotation from the list.
     The user interface changes to show commands applicable to drafting and annotation tasks.

4   To switch to the 3D Modeling workspace:
   • Click Workspace Switching.
   • Select 3D Modeling.

The user interface changes to show the commands applicable to 3D modeling tasks.

5   To return to the Civil 3D workspace:
    - Click Workspace Switching.
    - Select Civil 3D from the list.

The user interface changes to show the Civil 3D commands.

6   Click Workspace Switching. Select Workspace Settings from the list.

7   In the Workspace Settings dialog box, select Automatically Save Workspace Settings to enable the option. Click OK.

This setting ensures that changes you make to a workspace are automatically saved.

Create a Workspace

Next, you create and customize a workspace using the Customize User Interface (CUI) command.

1   Click Workspace Switching.

2   Select Save Current As from the list.

3   In the Save Workspace dialog box, for Name, enter **My Civil 3D Workspace**. Click Save.

4   Click Workspace Switching. Select Customize from the list.

5   In the Customize User Interface dialog box, under the Workspace collection, select the My Civil 3D Workspace (current) workspace. Note that the contents of the workspace are displayed in the Workspace Contents pane to the right.

6   In the Workspace Contents pane, notice that the workspace does not display any toolbars.

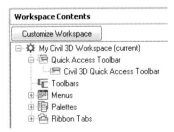

7   In the Workspace Contents pane, click Customize Workspace to enter Workspace Editing mode.

8   To add toolbars to the workspace, in the Customizations In All Files pane on the left side:
- Click to expand Partial Customization Files, ACAD, and Toolbars.
- Select the Draw and Modify check boxes. The toolbars are added to the My Civil 3D Workspace (current) workspace on the right side.
- Collapse Partial Customization Files.

9   To remove and move menus, in the Customizations In All Files pane:
- Click to expand Menus.
- At the bottom of the list, clear the Annotation and Inquiry check boxes.

10   In the Workspace Contents pane:
- Click to expand the Menus tree.
- Notice that the Annotation and Inquiry menus have been removed from the My Civil 3D Workspace (current) workspace.
- In the right pane, click and drag Lines/ Curves. Relocate it above General.
- Click Done to exit editing mode.

11   Click OK to close the Customize User Interface dialog box. Notice the display of the toolbars.

12   Next, you change how the menu bar is displayed:
- On the command line, enter **menubar**. Press ENTER.
- Enter **1** for Enter New Value for MENUBAR. Press ENTER.

The menu bar is displayed at the top of the screen. Notice that the Annotation and Inquiry menus are no longer visible. Also notice that the Lines/Curves menu is to the left of the General menu.

13   Finally, you confirm the effect of selecting Automatically Save Workspace Settings option:
- Click Workspace Switching.
- From the menu, select Civil 3D.
- Notice that the menu bar is no longer visible. Switch back to the My Civil 3D Workspace . Notice that the menu bar is visible. The visibility of the menu bar is controlled and saved with a workspace.

14   To turn off the menu bar in the My Civil 3D workspace:
- On the command line, enter **menubar**. Press ENTER.
- For New Value for MENUBAR, enter **0**. Press ENTER.

15   Close the drawing. Do not save the change.

Lesson 02 | Examining Toolspace

This lesson examines Toolspace. You use Toolspace to efficiently manage the data in drawings, as well as the settings, object styles, and label styles used to display that data.

The following illustration shows the Toolspace palette.

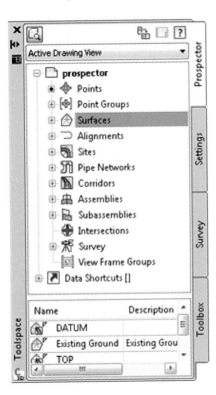

Objectives

After completing this lesson, you will be able to:

- Describe the function of Toolspace in drawing creation and management.

- List and describe the function of the four Toolspace components.

- Examine the two main components of Toolspace: the Prospector and Settings tabs.

About Toolspace

Toolspace is one of the primary interface components in Civil 3D. It provides an object-oriented view of the engineering data in your drawing and lists the object and label styles used to display the data. Toolspace is also used to display survey data and create external reports on your engineering data.

Definition of Toolspace

Toolspace is the primary tool that you use to control and display civil engineering data. With Toolspace, you can manage drawing and project data, create and manage settings and styles, manage survey data, and create reports.

Characteristics of Toolspace

Toolspace presents a large amount of data about the drawing, project, object styles, label styles, and drawing settings. The Toolspace palette can:

- Float or dock.

- Become semitransparent.

- Automatically hide itself.

- Be located on a second monitor.

Toolspace Components

There are two main components to Toolspace: the Prospector and Settings tabs. When you work with the Survey functionality, Toolspace displays a Survey tab. When you create reports, Toolspace displays a Toolbox tab.

Toolspace Tabs

The Toolspace tabs are described in the following table.

Prospector tab

The Prospector tab displays information about all of the Civil 3D objects in a drawing. Select this tab to manage drawing and project data. You work with point, point group, surface, alignment, profile, section, grading, parcel, and sheet layout data. In the bottom pane is the item view area. This pane shows additional information about the selected item. In this illustration, the item view displays a list of the surfaces in the drawing.

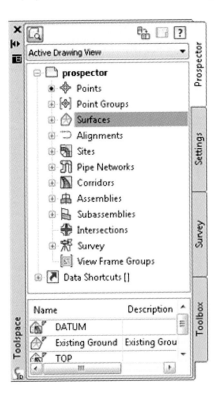

Settings tab

The Settings tab is where you manage object styles, label styles, and drawing settings for Civil 3D. Select this tab to configure drawings and drawing templates. You can specify drawing setup parameters such as units, scale, and coordinate zone. You can also set up object styles and object label styles.

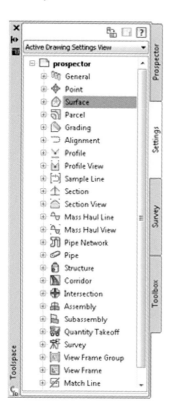

Survey tab

When you open the Survey Toolspace, this tab is added and used to manage survey observation data. You create survey databases, create survey networks, import survey data, and edit survey observations. You also create the survey network, points, and figures.

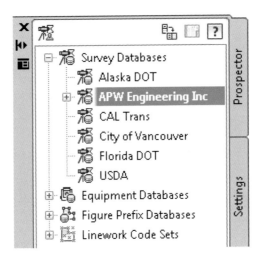

Toolbox tab

When you open the toolbox, this tab is added to create external reports on Civil 3D objects in a drawing.

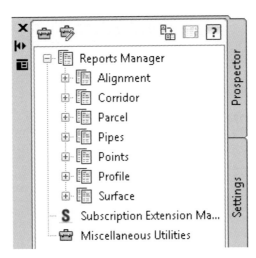

Exercise 01 | Examine Toolspace

In this exercise, you examine the two main components of Toolspace: the Prospector and Settings tabs.

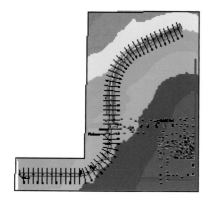

The completed exercise

Examine Prospector

First, you examine the Prospector tab.

1   Open…\AutoCAD Civil 3D Environment\prospector.dwg.

2   In Toolspace, Prospector tab, under the drawing name:

   • Click to expand Point Groups.

   • Select Additional TOPO Points.

   • In the Item View window, view the individual points.

   • Click to expand and resize the windows to see the point data.

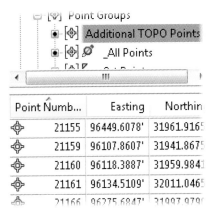

3   In the Item View area:

- Locate point number 20563.

- Right-click point 20563.

- Click Zoom To.

In the drawing area, notice that the drawing is zoomed to the point you selected.

20563
✕ 298.49
spot

4   Collapse Point Groups.

5   Click to expand Surfaces and Existing Ground to see the surface components.

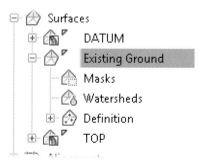

Each surface may have masks or watersheds, but each must be defined by one or more elements.

6   Click to expand Definition to view the elements that define the EG surface.

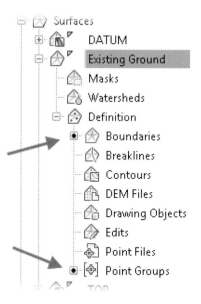

7   Right-click Existing Ground. Click Surface Properties.

You use the Surface Properties dialog box to view surface information and modify surface properties.

8   Click Cancel to close the Surface Properties dialog box.

9   In Prospector, click to expand Sites and Block Parcels. Click Block Parcels.

    You see the elements that the Block Parcels site contains.

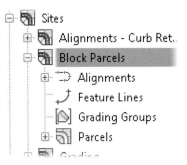

10   Click to expand Parcels. Select one of the parcels.

     You should see a preview of the parcel. If not, click the magnifying glass icon at the top
     of Prospector.

11   Right-click Parcels. Select Show Preview.

Examine Settings

Next, you examine the Settings tab.

1   In the Toolspace palette, click the Settings tab to view the collections of Civil 3D object and
    label styles.

2   Click to expand Point to view the settings and styles that you use to display point data
    in the drawing.

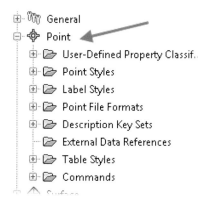

3   Click to expand Point Styles.

    The point style controls the display of the point node. The orange triangle next to the Basic
    style indicates that this style is currently in use.

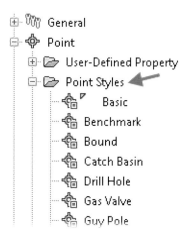

4   Collapse Point Styles and Expand Label Styles.

    Note that the Point#-Elevation-Description style is being referenced.

5   Click Prospector and do the following:

    • Click to expand Point Groups.

    • Right-click Additional TOPO Points.

    • Click Properties.

6   In the Point Group Properties dialog box, Information tab, under Default Styles:

- Note that the Basic point style and the Point#-Elevation-Description point label style are assigned to the Point Group.

- For Point label style, click Elevation Only.

- Click OK.

7   Zoom in to a point and view the point and label styles.

The labels on the points now display just the elevation.

8   Click the Settings tab and do the following:

- Collapse Point.

- Click to expand Surface and Surface Styles.

Note the surface styles in use. These are indicated with a yellow triangle.

9   Review the graphics window to see that the Existing Ground surface is displayed using the Contours 10' (Background) surface style.

10   Click Prospector. View the properties of the TOP and DATUM surfaces.

Note the application of the _No Display surface style.

11   In the drawing area, select any contour to select the surface. Right-click and then click Surface Properties.

12   In the Surface Properties dialog box, Information tab, do the following:

   • For Surface Style, click Elevation Banding (2D).

   • Click OK.

   Note the change in the drawing. The surface is now shaded based on elevation ranges.

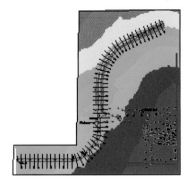

   • Close the drawing. Do not save the changes.

Lesson 03 | Creating Objects, Object Styles, and Label Styles

This lesson describes objects and how you create object and label styles.

Throughout the design process you create many Civil 3D objects with associated annotation. An object style is assigned to all Civil 3D objects and a label style is assigned to all Civil 3D labels. All objects are assigned a name. You can set default values for style assignment for both object styles and label styles. This way, when you create and annotate the design object, the object and associated label are automatically assigned the correct style.

You can also define object naming templates that automatically assign meaningful names to the design objects. For some objects you can assign default input parameters through the use of feature settings and command settings. All of this means less work for the designer when creating and annotating design objects in Civil 3D.

The following illustration shows Civil 3D point and parcel objects with their associated labels.

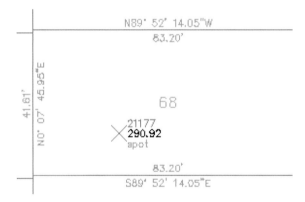

Objectives

After completing this lesson, you will be able to:

- Describe objects.

- Explain the purpose of object and label styles.

- Describe feature and command settings.

- Describe the process for creating object styles.

- Explain guidelines for creating label styles.

- Create an object style.

- Create a label style.

About Objects

Civil 3D is an object-oriented design environment that makes use of objects to display design components such as points, surfaces, alignments, profiles, pipe networks, and corridor models.

Most of the Civil 3D objects are shown in the illustration.

Civil 3D Objects	
⬦ Point	🔷 Point Groups
🔲 Site	🔲 Parcel
⬡ Surfaces	🔲 Grading Groups
↗ Feature Lines	↪ Alignment
⚊ Section	📊 Profile
🔳 Corridors	🔲 View Frame Groups
🔲 Assemblies	🔲 Subassemblies
🔲 Pipe Networks	🔲 Pipe Structure
⬟ Pipe	🔲 Survey
🔲 Networks	🔲 Figures

Definition of Objects

Objects are used to represent the survey, design, and construction elements you need to complete your site design or transportation design project. Objects have inherent properties that are used to control how the object is displayed and labeled, and how it functions.

Object Relationships

Objects have relationships that enable them to react to each other. These dynamic relationships are referred to as object reactivity. A few of these object relationships are highlighted in the following table.

Objects	Functional relationship
Points and Surfaces	Changing the elevation of point objects results in automatic updates to surface objects that reference the point objects.
Alignments and Surface Profiles	Changing the position or geometry of an alignment results in an automatic update to the surface profile objects.
Surfaces and Surface Profiles	Changing a surface results in an automatic update to surface profile objects that reference that surface.
Assemblies and Corridors	Changing the geometry of the assembly object (typical section) results in an automatic update to the corridor (road model) object.
Parcels and Parcel Segments	Bisecting an existing parcel object with another parcel segment or alignment object subdivides the original parcel object into two parcel objects.
Grading Object and Surfaces	Moving a grading object or changing a target surface results in automatic updates to the daylight line.

Object Naming Template and Input Parameters

This section describes the Object Naming Template and input parameters.

Object Naming Template

When objects are created they can be automatically assigned a name. You can use object naming templates to facilitate the assignment of meaningful names to objects. You can override the name assigned with the object naming template. Object naming templates are defined and saved in the company drawing template.

Input Parameters

For some objects you can assign default input parameters. For example, you can specify default elevations and descriptions for points, or an assembly insertion frequency for corridor models.

Examples of Objects

Objects are used to address all aspects of the planning, survey, and detailed design phases of site development and transportation projects. There are many other examples of objects including sample lines, sections, section views, grading, and tables. The following illustrations show additional examples of objects.

Survey Network

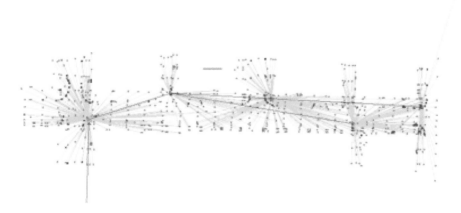

Survey network object

Profile

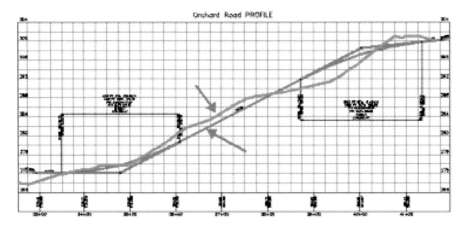

Profile view, surface profile, and layout profile objects

Corridor

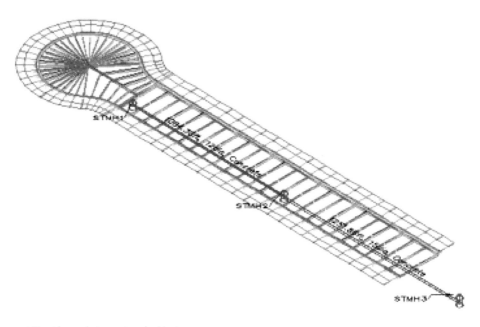

Corridor and pipe network objects

About Object and Label Styles

Every object in Civil 3D is displayed with an object style. Label styles are used to annotate the objects. When a design object is created, the assigned object and label styles are used to display the data according to the company or client standard.

Definition of Object Styles

Object styles are assigned to objects and control the display of objects. Every object in Civil 3D is displayed with an object style.

Definition of Label Styles

Label styles define the behavior, appearance, and content of labels. Label styles control the settings for the annotation and labeling of these objects.

Additional Information

- Object and label styles can assign the display properties (color, linetype, and lineweight) from directly within the style. This is referred to as By Style.

- Object and label styles can assign the display properties (color, linetype, and lineweight) from a layer. This is referred to as By Layer.

- All Civil 3D label styles have a dragged state display property. The dragged state display property controls the display of the label when it is dragged away from the object it is labeling.

- All Civil 3D labels scale with the drawing setup scale. This means that you do not need to create multiple sets of text to suit different scale requirements.

- All Civil 3D labels can be oriented to the object, view, or the world coordinate system.

Types of Label Styles

In the Toolspace Settings tree, the Label Styles collections contain one or more types of label styles that represent unique aspects of specific objects. You can create label styles for all objects, including points, surfaces, parcels, and alignments.

Points

You can create label styles for point objects to indicate information such as number, elevation, and description.

Surfaces

You can define label styles for surface objects such as slopes, spot elevation, watersheds, and contours.

Parcels

You can create label styles for parcel objects such as areas, lines, and curves.

Alignments

You can create label styles for alignment objects such as stations, station offsets, lines, curves, spirals, and tangent intersections.

Examples of Object Styles

The following illustrations show object style examples.

Parcel Style Example

The first example shows the style for a parcel property object. As you can see, the object style was created by Autodesk® and the design elements, such as Parcel Name Template, Boundary Pattern Fill Distance, and Observe Pattern Fill Distance, have defined values. Users can edit these design values as required.

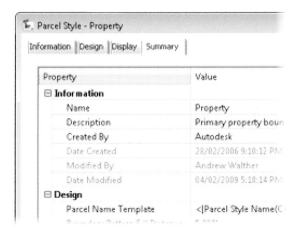

Grading Object Style

This next example shows the standard grading object style. The grading style specifies the marker and slope pattern styles that are used for grading objects. To distinguish between cut and fill grading, you apply different styles.

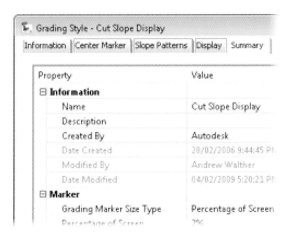

Label Style Example

The following illustration shows parcel objects with parcel number and parcel segment labels.

Settings to Configure Object and Label Styles

You use feature settings and command settings to configure object and label styles.

Feature Settings and Command Settings

- Feature Settings

  You use feature settings to change and view feature-specific settings, feature-specific ambient settings, and command-level settings. Examples include default styles, object naming templates, and input parameters.

- Command Settings

  Every Civil 3D command has corresponding command settings. Use command settings to change and view command-specific settings such as default styles, object naming templates, and input parameters.

Command Settings Example

Command and feature settings are modified from the Settings tab of Toolspace. The following illustration shows the command settings for the Create Profile View command.

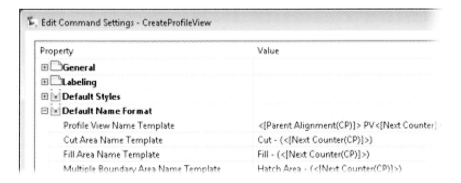

Creating Object Styles

Object styles, created by a CAD management role, are provided to users in a drawing template. Users create new drawings from the drawing template and choose from a selection of preconfigured object styles. Object styles should not be edited by users as part of their day-to-day work activity.

Creating Object Styles

The object style assigns display properties to object components. Every object has a different set of components. When you create an object style, you first select the components to be visible and then assign their display properties. You can control the display of the object components from directly within the object style, or by the layer.

Objects styles also incorporate other display settings that are specific to the object. For example, you can place different markers on the alignment geometry points in the alignment object style. With the surface object style, you have additional control over how the surface components are displayed. You can specify contour increments and smoothing factors, and you can assign a vertical exaggeration for all surface components to enhance 3D visualization.

Finally, all object styles have independent display properties for plan view, model view (3D), and section view. This means that the display of the object is a function of its view orientation.

Guidelines for Creating Label Styles

When you create an object you have the option to label the object. You can add or remove labels to objects any time. Users are typically provided with a selection of label styles that originate from a drawing template, and apply the label styles during their design activity.

Keep the following guidelines in mind when creating surface styles.

- As an alternative to working with layers, you can use No Display styles to suppress the display of an object by turning off the visibility of all object components.

- Use indicative naming conventions for object and label styles. The name of the style should indicate the function of the style.

Exercise 01 | Create an Object Style

In this exercise, you create a new object style to control the point display.

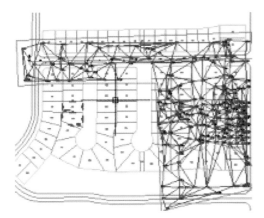

The completed exercise

1   Open…*AutoCAD Civil 3D Environment\object_styles.dwg*.

2   In Toolspace, on the Settings tab:

- Click to expand Surface, Surface Styles.

- Right-click Surface Styles.

- Click New.

3   In the Surface Style dialog box, Information tab:

- For Name, enter **Triangles**.

- For Description, enter **Display surface TIN lines**.

- Click Apply.

4   Click the Triangles tab and do the following:

- Notice that the surface style displays the triangles using the surface elevation. For a site with little relief, you may choose to exaggerate the elevations.

- For Triangle Display Mode, click Exaggerate Elevation.

- For Exaggerate Triangles by Scale Factor, enter **3**.

Triangle properties	Value
3D Geometry	
Triangle Display Mode	Exaggerate Elevation
Flatten Triangles to Elevation	0.000'
Exaggerate Triangles by Scale	3.000

5   Click the Display tab and do the following:

- Turn off the display of the Border component.

- Turn on the display of the Triangles component.

- Change the color of Triangles to blue (color 5).

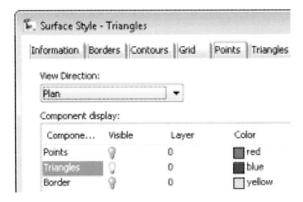

- Click the View Direction drop-down list.

Notice that the object style can change the display of the object differently depending on the view direction.

6   Next, you apply the new surface style.

Click the Summary tab and do the following:

- Click to expand Information. These properties can be modified by clicking in the Value cells.

- Click OK.

The new Triangles style is displayed in the Surface Styles list.

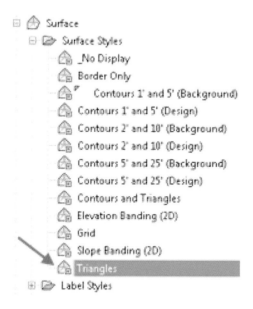

7   In Toolspace, on the Prospector:

- Click to expand Surfaces.

- Right-click Existing Ground.

- Click Properties.

8   In the Surface Properties dialog box, Information tab, for Surface style, click Triangles. Click OK.

The Existing Ground surface is now displayed with Triangles.

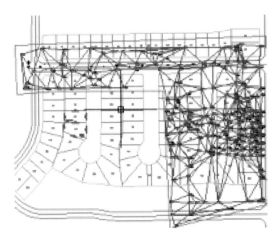

9   Close the drawing. Do not save the changes.

Exercise 02 | Create a Label Style

In this exercise, you create a label style and apply the new style to point objects.

The completed exercise

1 Open...*AutoCAD Civil 3D Environment\\label_styles.dwg*.

2 In Toolspace, on the Settings tab, click to expand Point, Label Styles.

   Notice that the active label style is Point#- Elevation-Description.

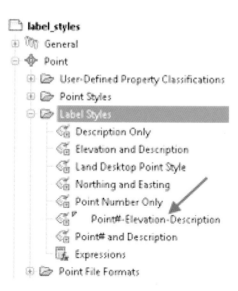

3 Right-click Point#-Elevation-Description. Click Copy to create a new style that is similar to this one.

4 In the Label Style Composer dialog box, Information tab:

   • For Name, enter **Elevation Only**.

   • Click Apply.

5   Click the General tab and do the following:

- In the Property column, under Label, for Visibility, select False from the list.

Notice the preview window changes by only showing the point marker with no label. All three label components are not visible. Also notice the layer on which the labels will be placed is V-NODE-TEXT. This setting is not desirable because you want the elevation label component to be visible.

- Return the value to True.

6   Click the Layout tab and do the following:

- Under Component Name, select Point Description from the list.

- For Visibility, click False.

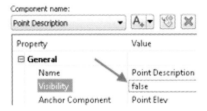

- Under Component Name, click Point Number.

- For Visibility, click False.

- Under Component Name, click Point Elev.

- For Text Height, enter **0.08**.

- For Color, click in the Value cell.

- Click the color palette.

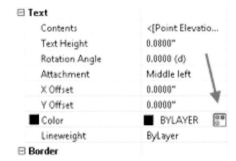

- Select Red (color 1) in the palette.

- Click OK.

The preview box shows a red elevation property.

7   Click the Summary tab and do the following:

- Review the properties for the components.

- Click OK to create the new label style.

Notice that the new style appears in the list of available label styles.

Next, you apply the new label style to point objects.

8   In Toolspace, on the Prospector:

- Click to expand Point Groups.

- Right-click Existing Topo.

- Click Properties.

9   In the Point Group Properties dialog box:

- Under Default Styles, for Point Label Style, click Elevation Only.

- Click OK.

10 Zoom in to a point to view the new label style.

11 Close the drawing. Do not save the changes.

Lesson 04 | Creating Drawing Templates

This lesson describes drawing templates (DWT) and the settings you can configure in the template. It also explains how you create drawing template files.

Drawing templates are among the most important elements to understand when creating and producing engineering and construction drawings. Configuring drawing templates based on your corporate or client's standards results in greater efficiency when creating drawings and sharing data. Standardized components are essential for consistency in engineering and construction drawings. Hence, all new drawings are created from drawing templates.

The following illustration shows the default Civil 3D templates.

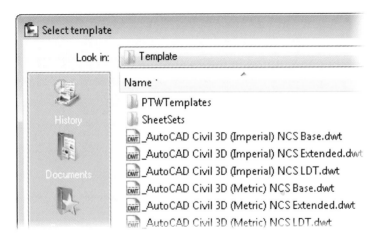

Objectives

After completing this lesson, you will be able to:

- Describe drawing templates.

- Describe template settings.

- List the steps in the process for creating drawing templates.

- Explore how you use command settings to set default styles, object naming templates, and default parameters for some objects.

- Create a drawing template based on an existing drawing's settings.

About Drawing Templates

Drawing templates are the starting point for your engineering and construction drawings. All new drawings must be created from drawing templates. To create a new drawing, you are prompted to select the drawing template to use. Your selection of available drawing templates should reflect either your corporate or your client standards.

Definition of Drawing Templates

Drawing templates are files that contain AutoCAD and Civil 3D standardized elements representing your company or client standards. All new drawings are created from drawing templates.

Example of Drawing Templates

Six preconfigured drawing templates are included in Civil 3D. Three are for imperial units, and three are for metric units. These drawing templates are all based on the National CAD Standard.

These templates can be used in the following circumstances:

- As a reference for learning how to create object and label styles.

- When organizations do not have their own established standards.

- Planning and analysis projects in which drawing production and presentation is not critical.

Template Settings

There are several template settings you can configure, both for AutoCAD elements and for Civil 3D elements.

AutoCAD Elements

Some of the AutoCAD standardized elements in a drawing template include:

- Layer names and visibility settings (color, linetype, lineweight, and so on).
- Block definitions.
- Text and dimension styles.
- Page setup definitions with viewports and title blocks.
- Miscellaneous system variables.

Civil 3D Elements

Some of the Civil 3D standardized elements in a drawing template include:

- Object and label styles.
- Drawing settings.
- Command and feature settings.

Drawing templates are preconfigured by a corporate CAD manager (or team) and should be centrally located on a network for all users to access. It is the role of the CAD manager to maintain the drawing templates, not that of the individual users.

Types of Template Settings

The following section describes the types of template settings.

Drawing Settings

You use drawing settings to specify the drawing units, scale, coordinate zone, transformation settings, object layers, and ambient settings. Drawing settings are stored in the drawing template, or DWT file. When you create a new drawing from a drawing template, the drawing settings are carried forward to the new drawing.

Transformation Settings

You use transformation settings to equate ground-level coordinates to grid-level coordinates. You specify grid and local coordinates for a base point and a rotation point. You also specify a scale factor. You can use label styles to label Civil 3D objects that reference either local coordinates or grid coordinates.

Ambient Settings

You use ambient settings to specify default units of measurement for Civil 3D objects.

AutoCAD Settings

AutoCAD settings that you configure include layers, text styles, dimension styles, symbols, system variables, and paper space layout definitions (including viewports and company title blocks).

Command and Feature Settings

Command and feature settings are used to assign the following:

- Default object and label styles.

- Object naming templates.

- Command specific settings.

Every feature category has command settings that can be found by expanding the Commands tree for the feature in the Settings tab of the Toolspace window. This is shown in the following illustration.

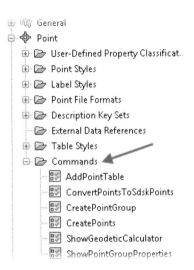

Command settings exist for every feature command. You modify command settings to assign default parameters such as styles, naming conventions, and algorithm parameters for objects. For example, the following illustration shows some of the command settings for the Create Profile From Surface command.

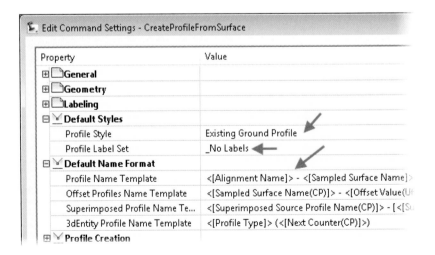

You can set command settings for all Civil 3D commands.

Creating Drawing Templates

Drawing templates are the basis for most new drawings. This section describes the process for creating drawing templates.

Process: Creating Drawing Templates

The following steps outline the process for creating drawing templates.

1   Make a backup copy of the drawing that will be the basis for your template. Open the drawing in Civil 3D.

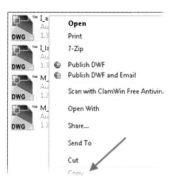

2   Erase all of the objects in the current drawing and purge undesired components.

3   Edit the drawing settings.

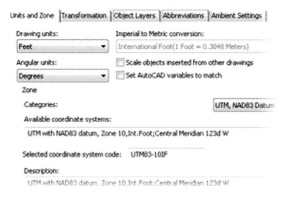

4   Save the drawing as a template.

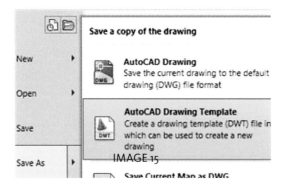

IMAGE 15

Exercise 01 | Explore Command Settings

In this exercise, you explore how you use the command settings to set default styles, object naming templates, and default parameters for some objects.

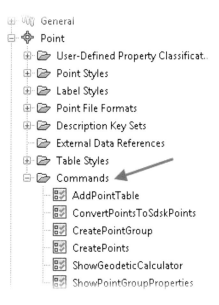

The completed exercise

1.   Open...*AutoCAD Civil 3D Environment\\command_settings.dwg.*

2.   In Toolspace, on the Settings tab, click to expand Alignment. Click to expand Commands.

You can specify defaults for object style assignment, object creation parameters, and naming conventions for each command in the Objects collection.

3.   Right-click CreateAlignmentLayout. Click Edit Command Settings.

4.   In the Edit Command Settings dialog box:

   • Click to expand Alignment Type Option.

   • Click to expand Default Styles.

Notice that the default alignment type is Centerline, and the default alignment style is Proposed. When new alignments are created using the Create by Layout command, these settings are used.

5   Under Default Styles:

  - For Alignment Style, click in the Value column. Click the ellipsis.

  - Select Layout from the list. Click OK.

  - For Alignment Label Set, click in the Value column. Click the ellipsis.

  - Select Major Minor and Geometry Points from the list. Click OK.

Notice the default point style has changed and a check mark has been placed in the override box to the right because this setting now overrides the settings at a higher level.

Alignment Type Opt		
Alignment Type	Centerline	
Default Styles		
Alignment Style	Layout	☑
Marker Style		☐
Alignment Label Set	Major Minor and Geometry Points	☑
Line Label Style	Bearing over Distance	☐
Curve Label Style	Delta over Length & Radius	☐

  - Click OK to close the dialog box.

6   To view the effects of the changes:

  - On the ribbon, click Alignment > Alignment Creation Tools.

  - In the Create Alignment dialog box, notice that the default alignment style is Layout, and the default alignment label set is Major Minor and Geometry Points.

  - Click Cancel.

7   In Settings, expand Surface, Commands.

8   Right-click AddContourLabeling. Click Edit Command Settings.

Note: You can also double-click AddContourLabeling to edit the command settings.

9   In the Edit Command Settings dialog box:

   • Click to expand Labeling. Notice that the Labeling Prompt Method is set to Command Line.

   • Click in the value cell to see the choices.

   You can control how you interact with Civil 3D through these settings. Many of these settings are the same within the Commands settings for the other Civil 3D features.

   • Click Cancel to close the Edit Command Settings dialog box.

10   In Settings, under Commands, right-click CreateSurface. Click Edit Command Settings.

11   In the Edit Command Settings dialog box:

   • Click to expand Default Styles.

   • Click to expand Build Options.

Property	Value
⊟ 📋 **Build Options**	
Copy Deleted Depend...	no
Exclude Elevations Le...	no
Elevation <	0.000'
Exclude Elevations Gr...	no
Elevation >	0.000'
Use Maximum Triang...	no
Maximum Triangle Le...	0.000'
Convert Proximity Bre...	yes
Allow Crossing Breakl...	no
Elevation to Use	Use first breakl...

These settings control various options used in building a surface. Notice that you can control elevations above or below given values by changing the No value to Yes, and then entering an elevation in the following value cell. Maximum Triangle Side Length can be set the same way. You can enable or disable crossing breaklines. These settings modify the algorithm to build surfaces.

- Click Cancel.

12   In Settings:

- Click to collapse Surface.

- Click to expand Profile, Commands.

- Double-click CreateProfileFromSurface.

13   In the Edit Command Settings dialog box:

- Click to expand Default Name Format.

- For Profile Name Template, click in the Value column. Click the ellipsis.

Notice that the name assigned to surface profiles calls the alignment name and the surface name.

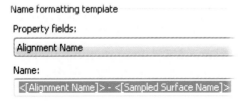

- Click Cancel twice.

14   Close the drawing. Do not save the changes.

Exercise 02 | Create a Drawing Template

In this exercise, you prepare a drawing and save it as a drawing template. Setting up a template drawing that contains all of the required settings, styles, layers, blocks, and parameters that can be used repetitively can save a lot of time and energy. If you want to keep the changes you make to a drawing's settings, you can save it as a template file.

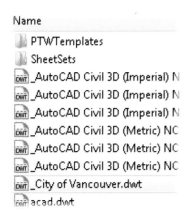

The completed exercise

1. Open ...\AutoCAD Civil 3D Environment\drawing_template.dwg.

2. Use Windows Explorer to make a backup copy of your current drawing.

   Next, you erase all of the objects in the current drawing and purge undesired blocks.

3. On the ribbon, click Layer Properties Manager.

   Do the following:

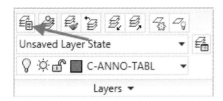

   - Ensure that all layers are on and thawed.

   - Close the Layer Properties Manager.

4   Zoom in to the extents of the drawing.

Note: You can also double-click the middle mouse button to achieve the same result.

5   To delete all of the points in the drawing:

- In Toolspace, on the Prospector, click to expand Point Groups.

- Right-click _All Points. Click Delete Points.

- Click OK to confirm the deletion.
  Note that the Existing Topo point group is out of date.

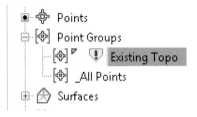

- Right-click Existing Topo. Click Update.

6   To delete the surface:

- Click to collapse Point Groups.

- Click to expand Surfaces.

- Right-click EG. Click Delete.

- Click Yes to confirm the deletion.

7   To delete the alignments:

- Click to collapse Surfaces. Click to expand Alignments.

- Click Centerline Alignments.

- In the item view area, press SHIFT and select all alignments.

- Right-click the selected alignments. Click Delete.

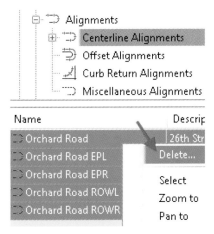

- Click Yes to confirm the deletion.

8   To delete the site and parcels in the site:

- Click to collapse Alignments. Click to expand Sites.

- Right-click Block Parcels. Click Delete.

- Click Yes to confirm the deletion.

All Civil 3D objects have now been deleted.

9   To erase the remaining entities in the drawing, you use the AutoCAD Erase command. At the command line, enter **Purge**. Press ENTER.

10 In the Purge dialog box, click to expand Blocks.

- Select FireHydrant. Right-click and click Purge.

- Click Purge to purge this item.

- Examine the layers listed in the Layers tree. Here, you purge the layers that you do not want in the template file.

- Click Close.

Next, you check the drawing settings.

11   In Settings, right-click the drawing name. Click Edit Drawing Settings.

12   In the Drawing Settings dialog box, on Units and Zone tab, for Scale select 1" = 100'.

13   On the Object Layers tab:

- Examine the layer names that are associated with the objects. Ensure that the names match your conventions.

- You can modify any object layer by clicking the layer cell for that object.

14   On the Ambient Settings tab:

- Click to expand Distance.

- For Precision, click 2.

- Click OK.

15   In the Settings dialog box:

- Click to expand Point, Label Styles.

- Right-click Land Desktop Point Style. Click Delete.

- Click Yes to confirm the deletion.

At this point, you can modify or delete a collection, or an individual style or setting as desired.

16   To save the template:

- Click the Application toolbar.

- Click Save As.

- Click AutoCAD Drawing Template.

17   In the Save Drawing As dialog box:

- For Name, enter **_City of Vancouver**.

- Click Save.

- Click OK.

The template drawing file can be used for any new drawing created while working for the city, meeting their settings and styles standards.

18   Click the Application toolbar. Click Close.

Next, you create a new drawing using the new template.

19   Click the Application toolbar. Click New.

20   In the Select Template dialog box, click _City of Vancouver.dwt. Click Open.

A new drawing is created that contains all of the settings and styles contained in the _City of Vancouver drawing template.

21   Close the drawing. Do not save the changes.

Lesson 05 | Creating Reports

This lesson describes how you create reports for design objects.

When you are working on projects, you often need to create reports for the design objects in your drawing. You can create reports for points, surfaces, alignments, profiles, pipes, and corridors. For each of these features there are several report formats to choose from.

The following illustration shows part of a Parcel Inverse report.

Parcel Inverse Report		Client: Client Company
Project Name: \Creating Parcel Tables and Reports\Maplewo		Project Description:
Report Date:		Prepared by: Preparer

Parcel SINGLE-FAMILY_ 601		
Point whose Northing is 78007.052 and whose Easting is 1030963.024		
	Bearing: N 4-34-43.362 E	Length: 51.935
Point whose Northing is 78058.822 and whose Easting is 1030967.170		
	Bearing: N 79-59-60.000 E	Length: 100.000
Point whose Northing is 78076.187 and whose Easting is 1031065.650		
	Curve	
	Direction P.C. to Radius:	S 59-13-37.803 W
	Radius Length:	945.585
	Delta:	05° 25' 04.74"
	Curve Length:	89.416
	Chord Length:	89.383
	Chord Direction:	S 28-3-49.830 E
	Direction Radius to P.T.:	N 64-38-42.538 E
Point whose Northing is 77997.313 and whose Easting is 1031107.701		
	Bearing: S 76-43-24.338 W	Length: 138.574
Point whose Northing is 77965.489 and whose Easting is 1030972.831		
	Bearing: N 13-16-35.662 W	Length: 42.704
Area		
	Square feet	11035.796

Objectives

After completing this lesson, you will be able to:

- Describe how you use the Reports Manager to create reports.

- Describe the process for creating reports.

- Use the Reports Manager to create reports for design objects.

About the Reports Manager

You use the Reports Manager to create reports in Civil 3D. When you open the Toolbox, a new tab is created on the Toolspace window. It shows the Reports Manager and a number of different reporting options.

Reports Manager

Definition of Reports Manager

The Reports Manager is the primary tool for creating reports in Civil 3D 2010. You use the Reports Manager to create different types of external reports for the Civil 3D features.

Examples

The following illustrations show examples of the types of reports you can create using the Reports Manager.

The first example shows a portion of a station and curve report for an alignment.

Curve Point Data

Description	Station	Northing	]
PC:	8+10.867	17831432.627	1695475.675
RP:		17831627.627	1695475.656
PT:	9+69.672	17831493.809	1695617.493

Circular Curve Data

Parameter	Value	Parameter	
Delta:	46° 39' 38.5329"	Type:	LEFT
Radius:	195.000	DOC:	29° 22' 56.82
Length:	158.805	Tangent:	84.103
Mid-Ord:	15.944	External:	17.364
Chord:	154.452	Course:	N 66° 39' 50.

Tangent Data

Description	PT Station	Northing	]
Start:	9+69.672	17831493.809	1695617.493

The following illustration shows a portion of a parcel area report.

Parcel Area Report
Report Date: 3/18/2008 11:58:57 AM

Parcel Name	Square Feet	Acres
Single-Family : 1	4927.657	0.113
Single-Family : 2	4137.530	0.095
Single-Family : 3	4137.530	0.095
Single-Family : 4	4137.530	0.095
Single-Family : 5	4137.530	0.095

Creating Reports

To create reports, you use the Toolbox, which is an additional tab on the Toolspace window. From the Toolbox, you can create different report types for different features and change the report settings.

The following illustration shows some of the report types that you can create.

Creating Reports

To create reports from within Civil 3D, use the Toolbox. The Toolbox appears as a tab on the Toolspace. You are presented with a number of different report types. Select the report type, and select the object(s) you wish to create reports for. Civil 3D then displays the report in Internet Explorer.

Guideline for Creating Reports

Keep the following guideline in mind when you create transportation assemblies.

- Format reports with your corporate or project information. Report settings you can change include the client and owner company name, contact, and other information. An example is shown in the following illustration.

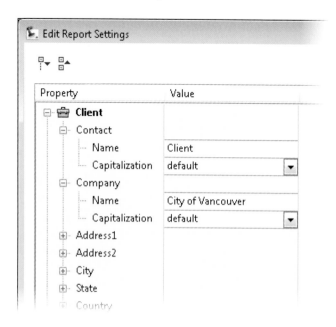

Exercise 01 | Create Reports

In this exercise, you create reports for design objects.

Alignment: 8th Avenue

Description:

Tangent Data	
Length:	259.884

Circular Curve Data	
Delta:	47° 02' 08.7649"
Radius:	570.866
Length:	468.641
Mid-Ord:	47.419
Chord:	455.591

The completed exercise

1 Open...\AutoCAD Civil 3D Environment\reports.dwg.

2 A Toolbox tab is displayed in Toolspace.

On the ribbon, Home tab, click Palettes menu. Click Toolbox.

3   In the Toolbox:

- Click to expand Reports Manager.

- Click to expand Points.

- Right-click Points_List. Click Execute.

4   In the Export to XML Report dialog box, click OK.

The report is displayed in a web browser.

Number	Northing (ft)	Easting (
35859	17831718.843	1696552.0
35858	17831765.807	1696458.4
21168	17831961.716	1696188.9
21563	17831867.344	1696443.8
36147	17831780.847	1696652.0
36148	17831783.713	1696630.5
36149	17831798.030	1696619.1
36150	17831790.402	1696600.3
36151	17831792.639	1696584.9
21860	17831876.173	1695756.1

5   Close the web browser.

Next, you create a report that shows the station and offset to points from an alignment.

6   In the Toolbox, right-click Station Offset to Points. Click Execute.

7   In the Create Reports - Station Offset to Points Report dialog box:

- Review the point list and review the alignments.

- Under Report Settings, select Orchard Road.

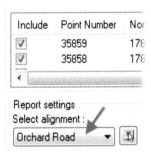

- Click Create Report.

8   The report is displayed in a web browser.

Alignment Name: Orchard Road
Description: 26th Street
Station Range: Start: 32+54.27,

Point	Station
35859	40+38.03
35858	39+44.93
21168	36+77.33
21563	39+31.28
36147	41+38.60
36148	41+17.17

- Close the web browser.

9   In the Create Reports - Station Offset to Points Report dialog box, click Done.

     Next, you create a parcel report.

10   In the Toolbox:

     • Click to collapse Points.

     • Click to expand Parcel. Notice the different reporting options for parcels.

     • Right-click Inverse_Report. Click Execute.

11   In the Export to XML Report dialog box:

     • Scroll down to see the parcels.

     • Click OK.

The report is displayed in a web browser. Notice the generic headers at the top of the report.
You can change how this information is displayed by changing the report settings.

Parcel Inverse Report

Project Name: C:\07-027 (Autodesk, AOTC
2009\Chapter 01 - Working in the Civil 3D Er
Reporting\reports.dwg

Report Date: 3/13/2008 5:22:54 PM

Parcel Single-Family : 1	
Point whose Northing is 17832074.468 and w	
	Bearing: ‹
Point whose Northing is 17831981.253 and w	
	Bearing: ‹
Point whose Northing is 17831980.813 and w	
	Bearing:]

12   In the Toolbox, click Report Settings.

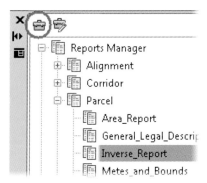

13   In the Edit Report Settings dialog box:

  • Click to expand Client.

  • Click to expand Company.

  • For Name, enter **City of Vancouver**.

  • Click to collapse the Client tree.

14   Click to expand Owner and Preparer:

  • For Name, enter **ABC Engineering**.

  • Click to expand the other trees and review the options.

  • Click OK.

  Next, you create an Alignment Report.

15   In the Toolbox:

  • Click to collapse Parcels.

  • Click to expand Alignment.

  • Right-click Alignment_Curve. Click Execute.

16   In the Export to XML Report dialog box, click Pick from Drawing.

17   In the drawing area, select 8th Avenue, the north-south running alignment on the west side of the site. Press ENTER.

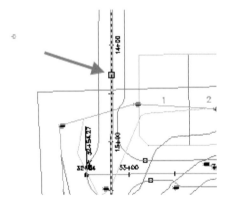

18   Click OK.

The report is displayed in the web browser. Notice the updated information for Client and Prepared By.

Alignment: 8th Avenue

Description:

	Tangent Data
Length:	259.884

	Circular Curve Data
Delta:	47° 02' 08.7649"
Radius:	570.866
Length:	468.641
Mid-Ord:	47.419
Chord:	455.591

19   Close the drawing. Do not save the changes.

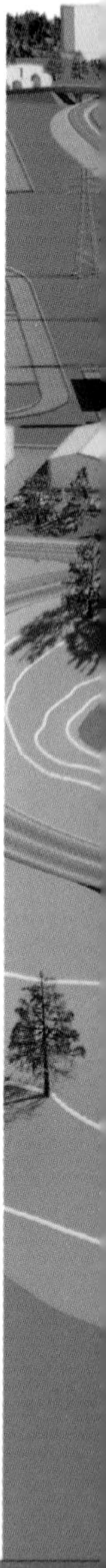

Chapter 02
Working with Survey

This chapter describes the Survey functionality in Civil 3D®. You can work directly with total station and GPS observation data to automatically create pre-engineering base plans and existing ground surface models. You can also edit field data to adjust control coordinates, backsight angles, prism heights and any other type of observed data. You use the Import Survey Data wizard to streamline the process for working with survey data.

The Creating Survey Databases lesson describes how you create and establish the units in the survey database. The Creating Survey Networks lesson describes how you use a survey network to organize data in the survey database. In the Creating Figure Styles and Prefixes lesson, you use figures to represent the base plan linework and surface breaklines. You also learn how to use figure prefixes to automate the assignment of the figure style. The Importing Survey Data lesson describes how you import survey field book files to the survey database. Finally, the Working with Survey Data lesson describes how you edit survey observation data and graphically represent this data as a survey network.

Objectives

After completing this chapter, you will be able to:

- Create and modify survey databases.

- Create survey networks.

- Create survey figure styles and survey figure prefixes.

- Import survey data from a field book file and begin base plan creation and surface modeling process.

- Check the survey database for errors and create survey points and figure objects.

Lesson 06 | Creating Survey Databases

This lesson describes how to create a local survey database, as well as how to modify the database settings.

Survey information is the essential starting point for any land development project. Before the planning, feasibility, permitting, or design process can begin, a land survey of the project site must be conducted. This land survey represents existing conditions and shows the site's boundaries, topography, infrastructure, utilities, and other critical features. Once collected, this data can be stored in a central repository, the survey database. The data in the survey database can be used by any user and reference any coordinate system.

Objectives

After completing this lesson, you will be able to:

- Describe survey databases.

- Define survey settings.

- Create a survey database and prepare it for importing survey data.

About Survey Databases

The survey database stores the necessary data for rendering objects based on survey measurements. By creating a survey database, you can use survey data collected from a number of sources. Survey observation data that resides in the survey database may be recreated in different drawings with different coordinate systems. This is especially useful when you need to produce drawings that represent grid coordinates, as well as drawings that represent ground coordinates from a common survey database.

Survey databases

Definition of the Survey Database

The survey database stores and manages survey-specific information for use in your projects. The database contains all the control points, known directions, observation measurements, traverse definitions, figures, and standard deviations based on equipment data. This information, which is entered in the Toolspace Survey tab, the survey editors, and the survey command window, includes observations imported from data collector files. You can store the survey database on a network server or on a local hard drive.

The following illustration shows survey information from a field book file.

Station:4, Backsight:3							
Number	Angle	Distance	Distance Ty...	Vertical	Vertical Ty...		Target H.
402	339.2051	148.783	Slope ▼	91.2824	Vertical Ar	▼	5.14]
403	335.5220	145.741	Slope ▼	91.2859	Vertical Ar	▼	5.14]
404	333.2854	148.369	Slope ▼	91.2940	Vertical Ar	▼	5.14]
405	332.5129	145.902	Slope ▼	91.2908	Vertical Ar	▼	5.14]
406	335.2525	143.015	Slope ▼	91.2740	Vertical Ar	▼	5.14]
407	336.4450	150.755	Slope ▼	91.3038	Vertical Ar	▼	5.14]
408	338.2405	148.422	Slope ▼	91.3139	Vertical Ar	▼	5.14]
409	340.2038	147.884	Slope ▼	91.3906	Vertical Ar	▼	5.1

Survey Database Characteristics

The survey database is intentionally kept separate and independent of your drawing project by the application for both practical and legal reasons:

- Original work done by registered surveyors is information that could have legal implications, and should not be altered without knowing the consequences.

- The survey data can be accessed through multiple drawings and can affect other objects, such as points and surfaces. For example, when you change a prism height or a backsight angle, associated point data automatically updates.

- Survey data is transformed according to the survey database coordinate system and the individual drawing coordinate system. If the drawing units and coordinate zone differ, then the survey is transformed.

Example

The following illustration shows a survey database and the related data sources.

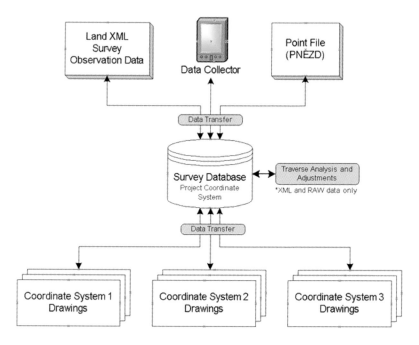

Survey Settings

You can configure two types of survey settings:

- Database

- User

Survey Database Settings

You use the following groups of survey database settings to control a number of survey database behaviors. You define the properties of the survey database by editing the values in the Survey Database Settings dialog box.

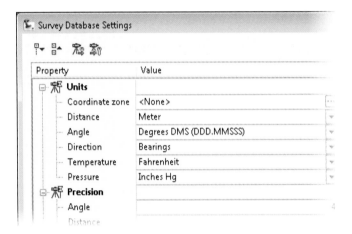

Survey Database Settings dialog box showing the survey database settings

Database Settings

Setting	Description
Units	Units settings select the Coordinate Zone, and Distance units, Angle units, Direction units, Temperature units, and Pressure units.
Precision	Precision settings control the precision for Angle, Distance, Elevation, Coordinate, and Latitude and Longitude. These settings are independent of the Drawing Settings precision settings.
Measurement Type Defaults	Establish default types for Angle, Distance, Vertical, and Target.
Measurement Corrections	Specify default measurement corrections applied to the survey observations.

Setting	Description
Traverse Analysis Defaults	Perform traverse analysis and determine the adjustment methods used for the analysis.
Least Squares Analysis Defaults	Specify the defaults for performing a least squares analysis on a network or a traverse.
Survey Command Window	Specify the survey database settings for interacting with the survey command window.
Point Protection	Specify the point protection settings applied for importing observations into a survey network.

Introduction to Modifying the Survey User Settings

You use the survey user settings to set the behaviors and characteristics of your sessions. These settings are specific to the Microsoft Windows user login, and they affect survey features, not database or drawing features.

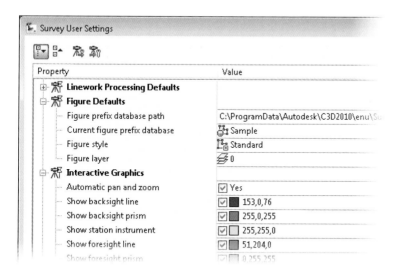

Survey User Settings dialog box

Survey User Settings

Setting	Value
Miscellaneous	Establish the external editor to use for displaying analysis input and output, and editing field book and batch file content.
Equipment Defaults	Establish the default equipment database settings including the database path, the current database, and the current equipment to use.
Figure Defaults	Establish equipment and figure prefix database information and figure style and layer defaults.
Interactive Graphics	Control the display of survey components during the import and entry of survey data.
Import Defaults	Specify the properties used when importing a field book or batch file into a selected survey network.
Export Defaults	Specify the properties used when exporting survey data to a field book file.
Network Preview	Specify the properties used when selecting a Network item in the Survey Toolspace tab.
Setup Preview	Specify the properties used when selecting a Setup item in the Survey Toolspace tab.
Figure Preview	Specify the properties used when selecting a Figure item in the Survey Toolspace tab.

Exercise 01 | Create a Survey Database

In this exercise, you create a survey database in Civil 3D and prepare it for importing survey data. The survey database is a separate file that contains the survey networks. Multiple users can access data in a survey database to create survey drawings.

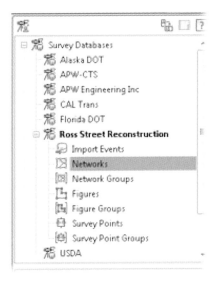

The completed exercise

1   Open...\Working with Survey\I_survey_database.dwg (M_survey_database.dwg).

2   On the ribbon, Home tab, click Survey Toolspace.

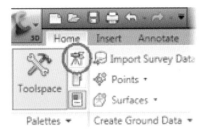

Note: If Survey Toolspace is open, clicking this icon closes the Survey Toolspace.

3 In Toolspace, Survey tab, right-click Survey Databases. Click New Local Survey Database.

4 In the New Local Survey Database dialog box, enter **Ross Street Reconstruction**. Click OK.

A survey database is added in Toolspace, Survey tab.

The survey database is physically represented with a folder structure on your hard drive. The default folder location is *C:\Civil 3D Projects* in your root drive. You may have changed this when installing the software.

Next, you change the units of measure in the survey database.

5 In Toolspace, on the Survey tab:

- Right-click Ross Street Reconstruction database.

- Click Edit Survey Database Settings.

6 In the Survey Database Settings dialog box:

- Click to expand Units.

- Change the Distance value to Meter.

Property	Value
⊟ 📷 **Units**	
Coordinate zone	<None>
Distance	International Foot
	International Foot
Angle	Meter
Direction	US Foot

The survey data that you import later has been collected in metric units. The survey database is therefore set up to represent metric units. Civil 3D performs a unit conversion by comparing the units of the drawing to those set in the survey database.

7   Click to expand the other items in the dialog box and review the settings. Click OK.

8   Close the drawing. Do not save the changes.

Lesson 07 | Creating Survey Networks

This lesson describes survey networks and how to create them. Survey networks are collections of survey control, instrument setup, and observation data, and are used to edit, organize and manage survey data.

Survey networks are typically created for site analysis, boundary survey, boundary analysis, topographic survey, and as-built survey phases of land development projects. Survey networks can also be used to organize survey data geographically.

The following illustration shows a survey network.

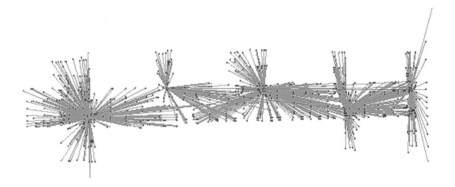

Objectives

After completing this lesson, you will be able to:

- Describe survey networks.

- List the components of a survey network.

- Create a survey network for use in a survey database.

About Survey Networks

Survey networks provide a repository for the data collected in land development projects. Each project phase, or location, can be represented by a network in the database. Once the network is created, survey data can be uploaded to the database from total station and GPS data files.

You can organize survey data by creating survey networks within the survey database for different stages of a land development project. Survey networks would be created for each of the following project phases:

- Site analysis

- Boundary survey

- Boundary analysis

- Topographic survey

Definition of Survey Networks

A survey network is a collection of related control points, instrument setups, survey observations, and defined traverses. Survey networks are stored in the survey database, and a single survey database can contain several survey networks.

Survey Network Basics

Each survey network is displayed in the survey database folder as a separate subfolder, and a survey database may contain several networks. You can import several survey data files, or raw data files, to a single survey network.

Survey networks exist in the survey database and can be inserted to and removed from any drawing connected to the survey database.

A survey network style is used to control the appearance of a survey network. You can use the survey network style to display control points, sideshot points, network lines, direction lines, sideshot lines, and error ellipses for traverses. You can create point objects and figure (linework) objects from the survey network after you check and adjust the network.

Survey networks contain the following data:

- Setups or stations.

- Control points and noncontrol points.

- Known directions.

- Observations.

- Traverses.

Example

The following illustration shows a survey network created in the Survey tab of Toolspace.

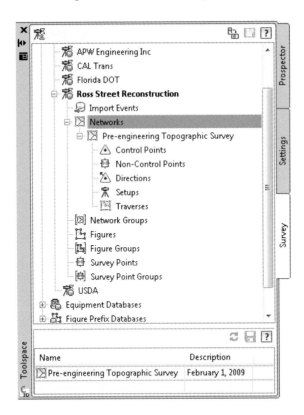

Survey Networks Components

A survey network contains information about a specific location such as a subdivision. This section lists some of the components of a survey network.

Survey Network Components

The following illustration shows the graphical components of a survey network.

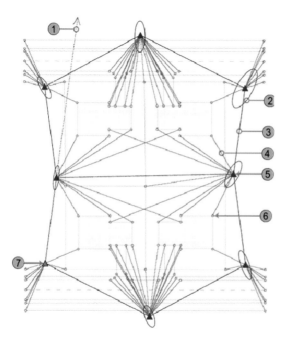

Graphical components of a survey network

(1) Direction line.　　(5) Unknown control point (that is, Setup).

(2) Error ellipse.　　(6) Sideshot point.

(3) Network line.　　(7) Known control point.

(4) Sideshot line.

Process: Creating Survey Networks

The following steps show how you create survey networks.

1. Create a new network in the survey database.

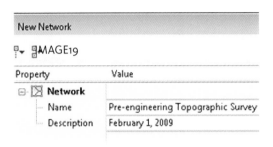

2. Modify the survey network setting.

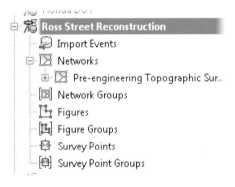

3. Display the survey network in a drawing.

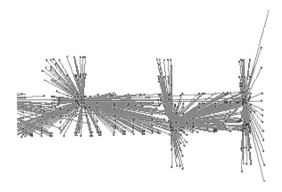

Exercise 01 | Create a Survey Network

In this exercise, you create a survey network in a survey database.

The following illustration shows a newly created survey network.

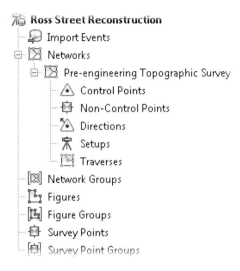

The completed exercise.

1   Open...\Working with Survey\I_survey_network.dwg (M_survey_network.dwg).

2   If the Survey tab is not visible in Toolspace, then on the ribbon, Home tab, click Survey Toolspace.

3   If the survey database does not exist in your drawing, complete the following steps. If it does exist, then start the exercise at step 5.

     In Toolspace, on the Survey tab:

     • Right-click Survey Databases. Click New Local Survey Database.

     • In the New Local Survey Database dialog box, enter **Ross Street Reconstruction**.

     • Click OK.

     The survey database is physically represented with a folder structure on your hard drive. The default folder location is *C:\Civil 3D Projects* in your root drive. You may have changed this when installing the software.

4   To change the units of measure in the survey database:

- In Toolspace, Survey tab, right-click Ross Street Reconstruction database.

- Click Edit Survey Database Settings.

- In the Survey Database Settings dialog box, expand Units.

- Change the Distance value to Meter.

- Click OK.

5   In Toolspace, if the Ross Street Reconstruction database is not open, right-click Ross Street Reconstruction. Click Open Survey Database.

6   Under the Ross Street Reconstruction database, right-click Networks. Click New.

7   In the New Network dialog box:

- For Name, enter **Pre-engineering Topographic Survey**.

- For Description, enter **February 1, 2009**.

- Click OK.

erty	Value
Network	
Name	Pre-engineering Topographic Survey
Description	February 1, 2009

8   In Toolspace, click to expand Pre-engineering Topographic Survey.

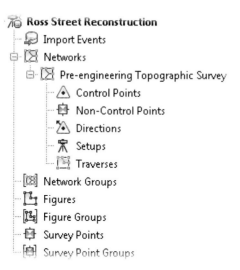

The survey network accommodates the following elements:

- Control points

- Noncontrol points

- Directions

- Setups

- Traverses

9   Close the drawing. Do not save the changes.

Lesson 08 | Creating Figure Styles and Prefixes

This lesson describes how you create survey figure styles and prefixes.

Survey figure prefixes help automate the creation of the pre-engineering base plan and the existing ground surface model. You use survey figures to represent pre-engineering base plan line work such as pavement edges, ditch bottoms, and road centerlines. They can also be used as breaklines for surface models. The survey figure style controls the display of the survey figure.

The following illustration shows a survey figure prefix with an assigned style.

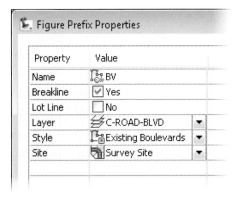

Objectives

After completing this lesson, you will be able to:

- Explain the function of survey figure styles and prefixes.

- Explain the key concepts related to the survey figure prefix database.

- Create survey figure styles and prefixes.

About Survey Figures and Prefixes

Survey figures and figure prefixes automate the production of the pre-engineering base plan and the existing ground surface model.

Many survey crews apply linework connectivity codes during their pre-engineering boundary and topographic surveys. Linework connectivity codes such as Begin, End, Continue, and C3 (to connect 3 points with an arc) result in the automatic creation of survey figures that represent ditch bottoms, road crowns, and sidewalk edges. The result is that the pre-engineering base plan and existing ground surface model is a direct by-product of the surveyor's field efforts.

Definition of Survey Figures

Survey figures are polylinear objects that contain line and arc segments. You use survey figures to connect similar survey features such as pavement edges, fence lines and sidewalks. Some survey figures can be used as breaklines for the surface model. Survey figure styles control the display of survey figures.

Definition of Survey Figure Prefixes

Survey figure prefixes automate the assignment of survey figure styles, which are used to control the display of the survey figures. Survey figure prefixes can also assign figures to layers and tag specific figures as breaklines. Survey figure prefixes are stored in an external file called the figure prefix database.

Survey Figure Examples

Examples of survey figures include road centerlines, edges of pavement, walls, buildings, property lines, streams, tops of banks, and ditches.

The following illustration shows the graphical components of a survey figure.

1 Figure line.

2 Vertex marker with marker style referencing a Multiview Block.

3 Midpoint markers placed at segment midpoints.

Process: Creating Survey Figure Styles and Prefixes

The following steps illustrate the process for creating survey figure styles and prefixes.

1   Determine survey figure style and display settings.

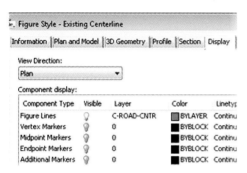

2   Create survey figure styles required by your client or organization.

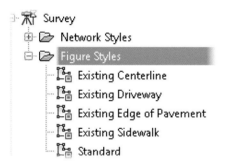

3   Create the survey figure prefix.

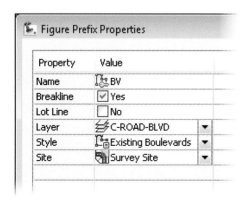

4   Create all required prefixes and assign styles to them.

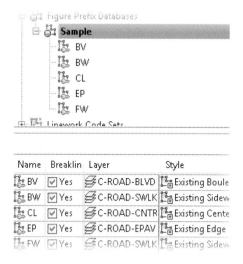

Survey Figure Prefix Database

Figure prefixes are stored in a figure prefix database. Prior to creating the figure prefixes you need to set the path and create a figure prefix database. The Figure prefix database should be centrally located locally on a computer network for all users to access.

Set Path

To set the path for the figure prefix database, you modify the survey user settings. To access survey user settings, click the Survey User Settings icon in Toolspace on the Survey tab.

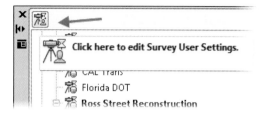

In the Survey User Settings dialog box, you set the path for the figure prefix database. A single figure prefix database should be created and centrally located on a server for all to access. You set the path by modifying the value for the figure prefix database path in figure defaults of survey user settings, as shown in the following illustration.

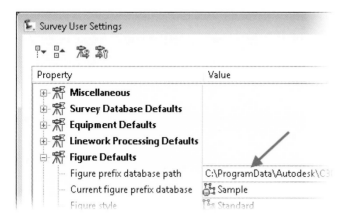

New Figure Prefix Database

You then create a new figure prefix database from Toolspace, on the Survey tab, as shown in the following illustration. A single figure prefix database is created and stored on the server for all users to share and access.

Exercise 01 | Create Figure Styles and Prefixes

In this exercise you create survey figure styles and prefixes. It is the responsibility of the CAD/Survey Manager to create the figure styles and the figure prefix database. The figure style meets the layer and appearance requirement of a client. Surveyors assign names to the figures when they are created in the field. When survey observation data is imported, the figure name is matched against the entries in the figure prefix database.

The completed exercise

Create Figure Styles

First, you create some figure styles.

1   Open...\Working with Survey\I_figure_styles.dwg (M_figure_styles.dwg).

2   In Toolspace, on the Settings tab:

   • Click to expand Survey.

   • Click to expand Figure Styles.

   • Right-click Figure Styles. Click New.

3   In the Figure Style dialog box, Information tab, for Name, enter **Existing Centerline**.

4   On the Display tab:

   • Click the light bulb to turn off all components except for the Figure Lines component.

   • For Figure Lines layer, click C-ROAD-CNTR.

   • For Figure Lines color, click BYLAYER.

   • Click OK.

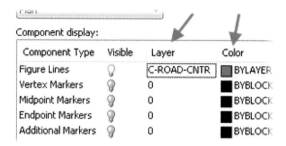

Component display:

Component Type	Visible	Layer	Color
Figure Lines	♀	C-ROAD-CNTR	■ BYLAYER
Vertex Markers	♀	0	■ BYBLOCK
Midpoint Markers	♀	0	■ BYBLOCK
Endpoint Markers	♀	0	■ BYBLOCK
Additional Markers	♀	0	■ BYBLOCK

5   Use the same procedure to create the remaining figure styles using the following data (Figure Name, Layer, Color):

- Existing Edge of Pavement, C-ROAD-EPAV, BYLAYER.

- Existing Sidewalk, C-ROAD-SWLK, BYLAYER.

- Existing Driveway, C-ROAD-DWAY, BYLAYER.

- Click OK.

You see the new figure styles in Toolspace.

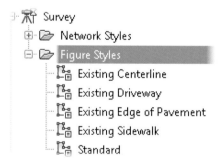

6   Close the drawing. Do not save the changes.

Create Figure Prefixes

Next, you create figure prefixes for the existing centerline, edges of pavement, sidewalks, and driveways. Surveyors assign names to figures when they are created in the field. When survey observation data is imported, the figure name is matched against the entries in the figure prefix database.

1   Open...*Working with Survey\I_figure_prefixes.dwg* (*M_figure_prefixes.dwg*).

2   If the Survey tab is not visible in Toolspace, then on the ribbon, Home tab, click Survey Toolspace.

3   In Toolspace, on the Survey tab:

   • Click to expand the Figure Prefix Databases.

   • Right-click Sample. Click New.

4   In the New Figure Prefix dialog box:

   • For Name, enter **CL**.

   • For Breakline, check Yes.

   • For Layer, change to C-ROAD-CNTR.

   • For Style, change to Existing Centerline.

   • Click OK.

Property	Value
Name	CL
Breakline	✓ Yes
Lot Line	☐ No
Layer	C-ROAD-CNTR
Style	Existing Centerline
Site	Survey Site

5   Create the remaining figure prefixes using the following data:

Name	Breakline	Layer	Style
EP	Yes	C-ROAD-EPAVE	Existing Edge of Pavement
FW	Yes	C-ROAD-SWLK	Existing Sidewalk
BW	Yes	C-ROAD-SWLK	Existing Sidewalk
DW	Yes	C-ROAD-DWAY	Existing Driveway
BV	Yes	C-ROAD-BLVD	Existing Boulevard

You are now finished with the figure prefix definitions. Notice that you can modify the figure prefix parameters directly from the item view on the Toolspace Survey tab.

6   Close the drawing. Do not save the changes.

Lesson 09 | Importing Survey Data

This lesson describes how to import survey data. First, it defines field book files, and then it describes a process for creating field book files. Finally, in the exercise, you review and import a field book file into the survey database.

Working with survey field book files enables you to edit and adjust the survey observation data directly within the Civil 3D environment. Any related data such as points, figures, and surfaces automatically update when you edit survey observation data. You import survey data to begin the base plan creation and surface modeling process.

Objectives

After completing this lesson, you will be able to:

- Describe survey field book files.

- Import field book files into a drawing.

- Review and import a field book file into a survey database.

About Field Book Files

Field book files are your primary source of survey data. This data is created from total station and GPS data files. Different brands of survey data collectors have their own observation data file format that can be converted to field book files, and then imported to Civil 3D.

The following illustration shows a survey field book file in a text editor.

```
Ross Street.fbk - Notepad
File  Edit  Format  View  Help
JOB Ross Street Avenue! DT02-01-2009 TM10:29:05
UNIT METER DMS
EDM OFFSET 0.0000
SCALEFACTOR 1
HORIZ ANGLE RIGHT
VERT ANGLE ZENITH
NEZ   1 5452310.3890 493970.3250 80.5050 "MONV4116UTM"
NEZ   2 5452304.8087 494174.6058 79.6490 "MONV654"
STN   1 1.800 "MONV4116UTM"
AZ    1 2 91.33531
BS    2 0.00000
! BS check 1-2: ZE90.18217,SD204.359,HD err=0.001,VD
PRISM  1.567
F1 VA 20 87.18413 15.666 92.02083 "POLER"
F1 VA 21 93.37010 16.634 91.52436 "TSR"
F1 VA 22 82.03279 18.155 91.50313 "LP"
F1 VA 23 81.28000 16.885 91.47238 "LP"
F1 VA 24 90.58058 21.247 91.45570 "TRR .2D"
BEG EW
F1 VA 25 83.41157 25.706 91.51512 "EW"
F1 VA 26 79.51421 25.943 91.47358 "EW"
```

Definition of Field Book Files

The field book (FBK) file is a survey data observation file. Field book files can contain pre-engineering and post construction survey data. Survey observation data includes the following:

- Control coordinates.

- Station setup information.

- Instrument and prism heights.

- Backsight angles.

- Sideshot data such as slope distance, horizontal angle, and vertical angle.

- Traverse data.

Importing the Field Book File

When you import a survey field book file to a survey network, the network in the survey database is populated with observation data such as control coordinates, prism heights, backsight angles, slope distances, horizontal angles, and vertical angles. The coordinates of the survey points and figures are also calculated.

Example of Field Book Files

The survey field book file contains observation data as shown in the following illustration:

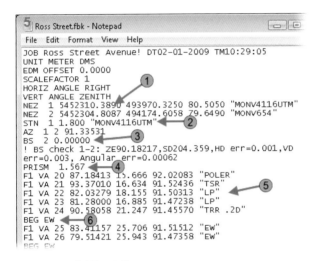

Survey field book file

①	NEZ - coordinates	Typically these are the points that the instrument is positioned over as well as the backsight point.
②	STN	Station setup command and height of instrument.
③	BS	Backsight angle.
④	PRISM	Rod height.
⑤	F1 VA	Sideshot observation data.
⑥	BEG	Begin statement to begin a figure. Other examples of linework connectivity codes include End, Continue, C3 (to connect three points with an arc), and Recall.

Importing Field Book Files

A field book file is a formatted file that supports observation data. You import the field book file to create observation data in the survey database. The different data collector manufacturers collect and store survey observation (raw) data in their own proprietary formats. You create field book files from the different observation data formats so you can import them to Civil 3D.

You can create field book files using the Survey Data Collection Link program found in the Survey menu as shown in the following illustration. You can also create field book files using the applications available from the survey data collector manufacturers.

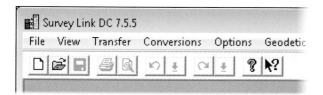

Importing Field Book Files

After you create the survey database and the survey network, you import the field book file to the survey network. You can import any number of field book files to a survey network.

When you import the field book file, the survey network in the survey database is populated with the survey observation data. You also have the option to create the following drawing objects:

- Survey network.

- Survey points.

- Survey figures.

Process: Importing Field Book Files

The following steps illustrate the process for importing field book files.

1   Import the field book file to the drawing.

Property	Value
Field book filename (.FBK)	C:\Users\Andrew
Current equipment database	Sample
Current equipment	Sample
Show interactive graphics	☑ Yes
Current figure prefix database	Sample
Process linework during import	☐ No
Current linework code set	Sample
Process linework sequence	By import order
Import event name	Ross Street.fbk
Import event description	
Assign offset to point identifiers	☐ No
Point identifier offset	
Insert network object	☑ Yes
Insert figure objects	☐ No
Insert survey points	☐ No
Display tolerance errors in Event Vie	☐ No

2   Review the survey data in the drawing.

Guidelines

Keep the following guidelines in mind when importing field book files.

- Surveyors must use standardized and consistent point descriptions in order to automate the generation of the pre-engineering base plan and existing ground surface model.

- Surveyors must apply linework connectivity codes in the field to automatically generate base plan linework.

- To help organize survey data, standards are required for layers, point styles, point label styles, description keys, and point group definitions.

- After creating the field book file, it is a good practice to review the contents prior to importing. You then import the field book file to the survey database and the drawing to begin the process for creating the existing base plan and surface model.

- A common practice is to initially create a survey network object when you import the field book file. This enables you to graphically visualize and assess whether or not traverse adjustments or edits are required. After you edit and adjust the survey data, you then create the survey point and figure objects from the survey database as a subsequent step.

Exercise 01 | Review and Import Field Book Files

In this exercise, you first review a field book file and then import it into the survey database.

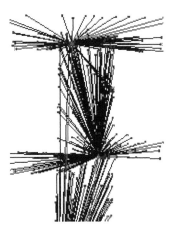

The completed exercise

1   Open...*Working with Survey\I_import_fieldbook.dwg* (*M_import_fieldbook.dwg*).

2   If you have created the figure prefix database, proceed to step 3. Otherwise, follow these steps:

   • On the Survey tab, click Edit Survey User Settings.

   • In the Survey Users Settings dialog box, for Figure Defaults, note the figure prefix database path. Click OK.

   • Using Windows Explorer, copy the provided figure prefix database, *Sample.fdb*, to the figure prefix database path location you just noted. Overwrite the existing file.

3   If the survey database exists, proceed to step 6.

   Otherwise, follow these steps:

   • On the Survey tab, right-click Survey Databases. Click New Local Survey Database.

   • In the New Local Survey Database dialog box, enter **Ross Street Reconstruction**.

   • Click OK.

The Survey Database is physically represented with a folder structure on your hard drive. The default folder location is *C:\Civil 3D Projects* in your root drive. You may have changed this when installing AutoCAD® Civil 3D®.

4   To change the units in the survey database:

- In Toolspace, on the Survey tab, right-click the Ross Street Reconstruction database. Click Edit Survey Database Settings.

- In the Database Settings dialog box, click to expand Units. For Distance, change the value to Meter.

- Click OK.

5   To create the survey network:

- On the Survey tab, for Ross Street Reconstruction, right-click Networks.

- Click New.

- In the New Network dialog box, for Name enter **Pre-engineering Topographic Survey**.

- For Description, enter **February 01, 2009**.

- Click OK.

6   On the Survey tab, click to expand Survey Databases. Right-click Ross Street Reconstruction. Click Open Survey Database.

Next, you review the field book file.

7   For Ross Street Reconstruction, right-click Pre- engineering Topographic Survey. Click Edit Field Book.

8   In the Field Book Filename (.FBK) dialog box:

- Browse to where you installed your Civil 3D Essentials 2010 datasets.

- Open Ross *Street.fbk* (*\Autodesk Learning\AutoCAD Civil 3D 2010\Learning\Working with Survey\Ross Street.fbk*).

- You may be prompted to select a text editor.

An alternate method is to browse to the folder and open *Ross Street.fbk* with your favorite text editor.

```
[OB Ross Street Avenue! DT02-01-2009 TM10:
UNIT METER DMS
EDM OFFSET 0.0000
SCALEFACTOR 1
HORIZ ANGLE RIGHT
VERT ANGLE ZENITH
NEZ  1 5452310.3890 493970.3250 80.5050 "M
NEZ  2 5452304.8087 494174.6058 79.6490 "M
STN  1 1.800 "MONV4116UTM"
AZ   1 2 91.33531
BS   2 0.00000
```

9   Review the contents of the field book file. Note the linework connectivity codes such as BEG, CONT, and END. Close the text editor when you are finished.

Next, you import the field book file.

10 On the Survey tab, right-click Pre-engineering Topographic Survey network. Click Import > Import Field Book.

11 In the Field Book Filename (.FBK) dialog box:

   • Browse to where you installed your Civil 3D Essentials 2010 datasets.

   • Open *Ross Street.fbk* (\Autodesk Learning\AutoCAD Civil 3D 2010\Learning\Working with Survey\Ross Street.fbk).

12 In the Import Field Book dialog box:

   • Select the Show Interactive Graphics check box.

   • Clear Process Linework During Import.

   • Select the Insert Network Object check box.

   • Click OK.

13 Watch as the survey data is imported to the drawing. This process takes a few minutes. The survey data is represented with a survey network.

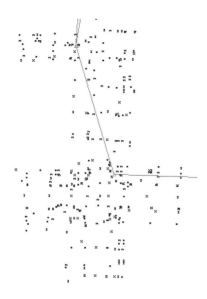

Next, you modify the survey network style to show the sideshot lines.

14 In the drawing area, click the survey network. Right-click and click Edit Survey Network Style.

15 In the Network Style - Basic dialog box, Display tab, for Sideshot Lines, click the lightbulb to turn on the component. Click OK.

Review the changes in the drawing area.

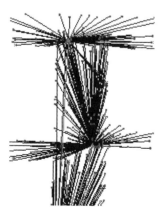

16 Close the drawing. Do not save the changes.

Lesson 10 | Working with Survey Data

This lesson describes how to work with survey data. It defines survey data and figures, and describes a process for creating points and figures. In the exercise, you correct some errors in the survey database, and create survey points and figure objects in the drawing from the survey data.

You can edit and perform traverse adjustments on survey data. Once the edits and adjustments are complete, you create the survey points and survey figure objects in the drawing directly from the data in the database. When you adjust and edit the survey network data again, the survey drawing objects automatically update.

Objectives

After completing this lesson, you will be able to:

- Describe survey observation data.

- Describe the process for creating survey points and figures.

- Edit the survey observation data to correct field errors.

About Survey Observation Data

Survey observation data is the basis for your design. This section defines survey data and figures, and provides examples of each.

Definition of Survey Observation Data

Survey observation data is the observed data that is recorded during a field survey, and usually consists of measured horizontal angles, vertical angles, and slope distances. Survey observation data is sometimes referred to as raw survey data. Observation data from a total station survey includes the following:

- Control coordinates.

- Station setup information.

- Instrument and prism heights.

- Backsight angles.

- Sideshot data such as slope distance, horizontal angle, and vertical angle.

- Traverse data.

Definition of Figures

Figures represent base plan line work such as pavement edges, centerlines, gutter lines, and sidewalks. In the field, surveyors apply figure connectivity codes, such as Begin and End, to automatically generate the base plan figures. Surveyors assign names to figures using the Begin command. Figures can be converted to breaklines for surface modeling. Figure display is controlled with a figure style. Figures are organized in figure groups.

Survey Data Characteristics

Survey network, point, and figure data are stored in the survey database and can be removed from or inserted into the drawing at any time. When you make changes to survey observation data, all associated data automatically updates.

For instance, if you change the prism height for a series of observed points, the observed point elevations update. Similarly, if you change the coordinate of a control point, the coordinates of the points observed (surveyed) from that control point update.

Observation Data Example

Survey observation data is displayed in the Survey Network on the Survey Toolspace. After you import the field book file, you can expand the trees in the Networks collection to see observation data. When you edit the observation data you change the contents of the Survey database. You then recalculate the Survey Network.

The following illustration shows the control points, directions, and setups in a survey network.

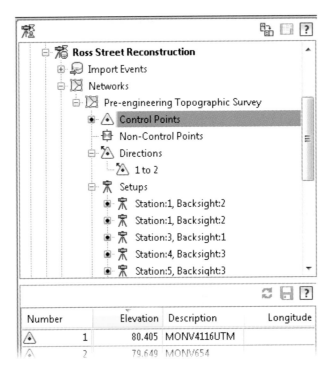

Survey observation data also consists of all of the observed figures and points as shown in the following illustration.

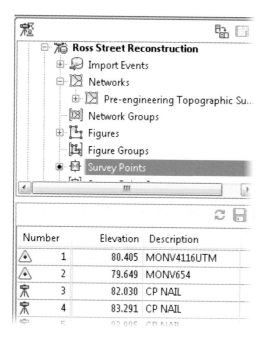

Creating Points and Figures

After you have imported your survey data you can then create survey points and figures.

Creating Points and Figures

After importing the survey network to the drawing so you can visualize, check, and adjust your survey, you then remove the network from the drawing and create survey points in the drawing.

Survey points created after removing the network

The next step is to insert the survey figures to the drawing area. The addition of layer linework completes the creation of the pre-engineering base plan.

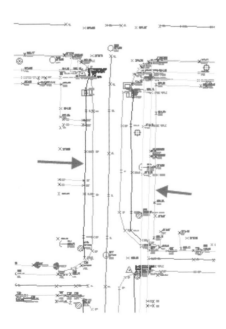

Guidelines for Creating Points and Figures

Keep the following guidelines in mind when you create points and figures.

- You can use survey observation data from a field book file, a Land XML file, or from manual input methods.

- To ensure accuracy survey observation data should only be edited by surveyors.

- You can use automatic recalculation for the survey network and the figures. Otherwise when you make changes to the survey observation data, you need to recalculate the survey network and figures as individual steps.

Exercise 01 | Work with Survey Data

In this exercise you edit the observation data to correct for field errors. Part of the survey data reduction process is to check the observation data for errors. If necessary you could also apply a traverse adjustment to the survey network.

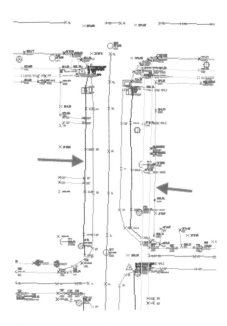

The completed exercise

Import Survey Data

1   Open *Working with Survey\I_edit_observation_data.dwg* (*M_edit_observation_data.dwg*).

2   If you have created the figure prefix database, proceed to step 3. If not, follow these steps:

   • On the Survey tab, click Edit Survey User Settings.

   • In the Survey Users Settings dialog box, for Figure Defaults, note the Figure Prefix Database Path. Click OK.

   • Using Windows Explorer, copy the provided figure prefix database, Sample.fdb, to the figure prefix database path location you just noted. Overwrite the existing file.

3   If the survey database exists, proceed to step 6. If it does not, follow these steps:

- On the Survey tab, right-click Survey Databases. Click New Local Survey Database.

- In the New Local Survey Database dialog box, for Name, enter **Ross Street Reconstruction**. Click OK.

The Survey Database is physically represented with a folder structure on your hard drive. The default folder location is *C:\Civil 3D Projects* in your root drive. You may have changed this when installing Civil 3D.

4   Next, change the units in the survey database:

- On the Survey tab, right-click Ross Street Reconstruction database. Click Edit Survey Database Settings.

- In the Survey Database Settings dialog box, click to expand Units.

- For Distance, change the value to Meter.

- Click OK.

5   Now create the survey network:

- On the Survey tab, for Ross Street Reconstruction, right-click Networks. Click New.

- In the New Network dialog box, for Name, enter **Pre-engineering Topographic Survey**.

- For Description, enter **February 01, 2009**.

- Click OK.

6   To open the survey database, right-click Ross Street Reconstruction. Click Open Survey Database.

7   If you have imported the field book file, proceed to the next exercise. Otherwise, follow these steps:

- On the Survey tab, right-click Pre- engineering Topographic Survey network.

- Click Import > Import Field Book.

- In the Field Book Filename (.FBK) dialog box, browse to where you installed your Civil 3D Essentials 2010 datasets.

- Select *Ross Street.fbk* (*\Autodesk Learning\AutoCAD Civil 3D 2010\Learning\Working with Survey\Ross Street.fbk*).

- Click Open.

8   In the Import Field Book dialog box set the properties as shown in the following illustration.

Field book filename (.FBK)	C:\Users\Andrew
Current equipment database	🗃 Sample
Current equipment	🖥 Sample
Show interactive graphics	☑ Yes
Current figure prefix database	🗂 Sample
Process linework during import	☐ No
Current linework code set	Sample
Process linework sequence	By import order
Import event name	Ross Street.fbk
Import event description	
Assign offset to point identifiers	☐ No
Point identifier offset	
Insert network object	☑ Yes
Insert figure objects	☐ No
Insert survey points	☐ No
Display tolerance errors in Event View	☐ No

- Click OK.

9   Watch as the survey data is imported to the drawing. This process takes a few minutes. The survey data is represented with a survey network.

Edit Observation Data

Notice that you can change properties for the control points and other survey observation data. Doing so results in updates to the survey network, point, and figures. All other dependent data are also updated.

1   Begin by noting the elevation of an observed point. After editing the control point data you notice that the elevation of the dependent points changes.

In Toolspace, on the Survey tab, right-click Ross Street Reconstruction. Click Open.

2   To insert the survey network to the drawing:

- Right-click Pre-Engineering Topographic Survey. Click Insert to Drawing.

- Click Survey Points.

- Notice that the elevation of point number 20 is 80.182.

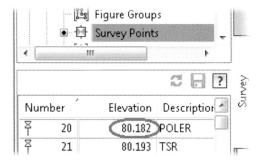

3   To correct an error with a control point elevation:

- On the Survey tab, expand Networks, Pre- engineering Topographic Survey.

- Click Control Points.

- In the item view area, for control point number 1, for Elevation, change the value to **80.405**.

- Press ENTER.

- Click Save.

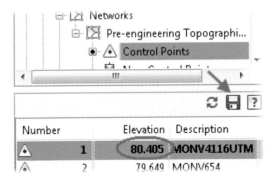

4   Notice the exclamation mark next to the Pre- engineering Topographic Survey network. The survey network is out of date and needs to be updated.

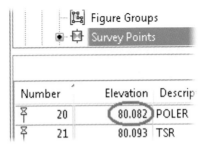

5   Right-click Pre-engineering Topographic Survey. Click Update Network.

6   Click Survey Points. Review the elevation for point number 20.

    The elevation has changed from 80.182 to 80.082 to reflect the adjustment of the elevation for control point number 1.

7   In the database a prism (target) height was entered incorrectly for the Station:3, Backsight:1 setup and must be corrected. To adjust the prism heights:

- On the Survey tab, Pre-engineering Topographic Survey network, click to expand Setups.

- Notice the different instrument setups.

- Right-click Setup Station:3, Backsight:1.Click Show Properties to review the setup properties.

- Right-click Setup Station:3, Backsight:1, click Edit Observations.

The observations (sideshots) from this setup are graphically highlighted.

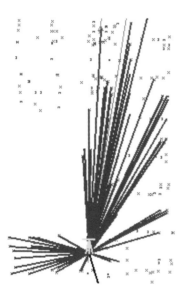

8   The Panorama window appears showing the Observations Editor. To review the observation data:

- Scroll down to point numbers 355, 356 and 357. Note the target height of 2.650. This target height was entered incorrectly.

- Scroll to the right and note the elevations of these three points: Point 355 - 82.015 Point 356 - 82.150 Point 357 - 82.243.

9   Next, you change the target height for the three points. In Panorama:

- Scroll back to the left.

- CTRL+click the three points (355, 356, and 357).

- Right-click the Target Height column header. Click Edit.

- Change the value to **2.55**.

- Press ENTER. Civil 3D recalculates the observations. Notice that the data is now bolded.

- Click Save. The data is no longer bolded.

This indicates that the changes have been saved.

- Close Panorama.

10 In Toolspace, on the Survey tab, notice that the Pre-engineering Topographic Survey network is now out of date. To update the network:

- Right-click Pre-engineering Topographic Survey. Click Update Network. It may take a few seconds.

- Click on Survey Points.

- In the item view area, scroll down to point numbers 355, 356, and 357.

- Notice that the elevations of the point data have updated based on the change to the target height.

11 The Figures data is now out of date. The figure vertices contain elevation data, which may have changed when the prism heights were updated. To update Figures, do the following:

- Right-click Figures.

- Click Update Figures.

Data residing in the survey database has been changed. The survey network and figure data have been updated.

Create Points and Figures

After the surveyor has checked and adjusted the survey network, the next step is to create the survey points and figure objects in the drawing. You also remove the survey network from the drawing to complete the exercise.

1 To add the survey points to the drawing:

- On the Survey tab, Ross Road Reconstruction database, right-click Survey Points.

- Click Points > Insert Into Drawing.

The drawing is updated with point objects that can be used for surface modeling and other design tasks.

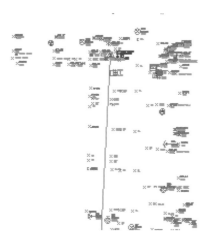

2   To add the figure objects to the drawing:

- On the Survey tab, right-click Figures.

- Click Insert Into Drawing.

The drawing is updated with the figures that represent the base plan linework.

3   To remove the survey network from the drawing:

- In Toolspace, Survey tab, right-click Pre- engineering Topographic Survey network.

- Click Remove from Drawing.

- In the warning message dialog box, click Yes.

Graphically navigate the drawing and review the data.

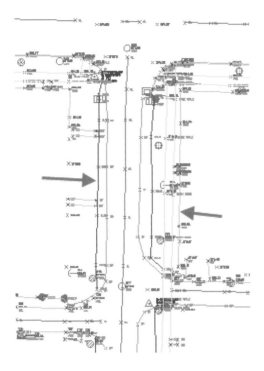

4   Close the drawing. Do not save the changes.

Chapter 03
Points

This chapter describes how you work with points. Points are one of the most common elements in the design environment. In the Importing and Creating Points lesson you import points from an external text file to the drawing. In the Managing Points lesson you learn how to work with description keys to assign point styles, symbols, and label styles. You use point groups to organize and manage points.

▶ **Objectives**

After completing this chapter, you will be able to:

- Import and create points.

- Manage points using description keys, point groups, and point tables.

Lesson 11 | Importing and Creating Points

This lesson describes how to import and create points. It describes points and how you can use them in different phases of an engineering project.

Points are one of the most fundamental elements in site development and transportation projects. They can be used to represent existing conditions and proposed construction locations. Point data that represents existing conditions is usually created from survey data files. Points used for construction are created by engineers and construction personnel. Once a design is completed, engineers extract point data from the design, and then send this information to the field for construction staking.

Objectives

After completing this lesson, you will be able to:

- Explain how points are used in different phases of an engineering project.

- Describe how to use grips to modify points and point labels.

- Describe which objects can be used to create points in a drawing.

- Import points from an external text file into a drawing.

About Points

You create points from survey data to generate a base plan drawing and a surface object to model the existing conditions. You also create points from Civil 3D® object data that can be used for construction staking.

Definition of Points

A point represents a singular location in space with elevation, northing, and easting coordinates. It also has properties that represent the point number, a raw (field) description, and an expanded (full) description. For modeling existing conditions, points are typically imported from a text file or created from the survey database.

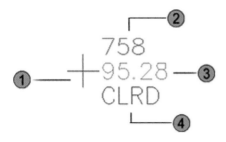

Components of a point

① Point marker

② Point number

③ Point elevation

④ Point description

Use of Points

Points are used in many different phases of an engineering project.

- **Legal Survey and Survey Control**

Before the engineering begins, legal surveyors use points to define land parcel boundaries. Points represent the corners of property boundaries and are then connected with parcel segments to define the land parcel. During this process, surveyors also use points to represent horizontal and vertical survey control points. These points are referenced when establishing locations for boundary, topographic, and construction surveys.

- **Topographic Survey**

Topographic surveyors use points to define existing conditions prior to design and construction. Existing conditions are represented with a pre-engineering base plan drawing, and an existing ground surface model. The base plan drawing is created with symbols and linework. Symbols are inserted at spot locations such as catch basins, manholes, and power poles. Base plan linework is created by connecting points with similar descriptions, such as edge of pavement and bottom of ditch. The existing ground surface model is created from points that represent the true topography. Points representing manhole inverts and fire hydrant tops are excluded from the existing ground surface model.

- **Design and Construction**

Designers create points on the design model that can be used for construction staking. Points created on the design model contain description, coordinate, and elevation data that contractors can use to lay out the proposed design for field construction.

Examples

The following are examples of how points are used in projects.

Points are imported from a text file that usually contains point number, northing, easting, elevation, and description. If point numbers are not provided in the text file, they are automatically created during the import. The following illustration shows an external text file with point data.

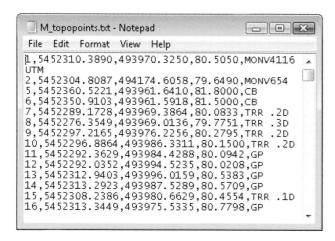

Prior to developing a parcel of land, a boundary survey is completed to define the limits of the parcel boundary. Surveyors use points to show property corners and survey control locations as shown in the following illustration.

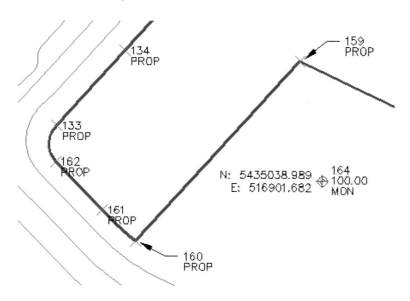

Points are also used for pre-engineering topographic mapping and the creation of the existing ground surface model as shown in the following illustration.

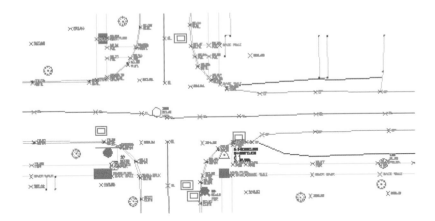

Points can also be used for construction purposes. Construction points can be created for road gutter lines and uploaded to a data collector for construction staking as shown in the following illustration.

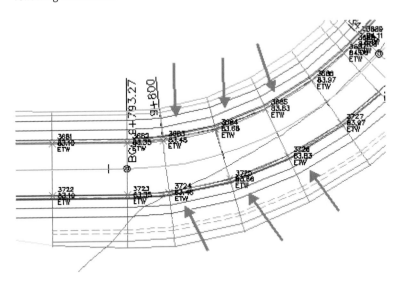

About Grips

Often points are created in close proximity to each other. This can make it difficult to read the data associated with the points. You can use grips to control the positioning of labels and other information associated with points.

Definition of Grips

Grips are small figures that appear at strategic points when you select an object.

Grips are not displayed on objects that are on locked layers.

Grip Types

There are four types of grips: drag point, drag label, toggle sub-item, and rotate point. The following illustration shows the grips used to control the display of points and points labels.

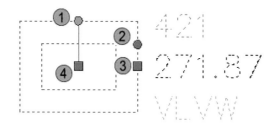

(1) Rotate Point Grip

Use the green circular rotate point grip to rotate the marker about its insertion point. You can use this grip to orient symbols to properly align with base plan linework.

(2) Toggle Sub-Item Grip

Use the toggle sub-item grip to display grips for the individual point components. You can then move the point components individually.

(3) Drag Label Grip

Use the drag label grip to reposition the entire point label away from the point object. When you use the drag label grip, the point label adopts its dragged state display property.

(4) Drag Point Grip

The drag point grip is the square grip that is displayed on the point marker. When you click the drag point grip you can move unlocked points to a new location. The northing and easting values for the point are updated based on the new location.

Grips Examples

The following illustration shows how an object is affected when the rotate point grip is used.

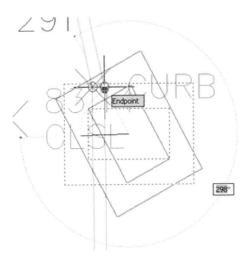

In the next illustration, the drag label grip is used to relocate a label.

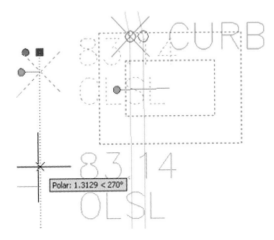

Objects Used to Create Points

You can create points by using objects in your drawing as a reference. For example, you can create points from the following:

- Parcels
- Surfaces
- Alignments

The following illustration shows the Create Points toolbar.

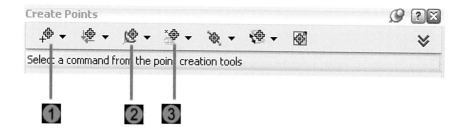

① Create points manually and other creation methods.

② Create points from horizontal and vertical alignments.

③ Create points on surfaces and polylines.

Point Data Created with Parcels and Surfaces

Creating points on a parcel object generates points that represent the endpoints of a lot line. You create points from a surface when you need to identify or isolate specific locations on the surface. Each point is assigned an elevation from a location on the surface.

You may need to use a surface to create points if the design requires you to mark spot elevations on your finished plan. You can add the points as needed and label and manage them as a group. You can also create random points when you need existing ground spot elevations to label a topographic survey plan, or finished ground spot elevations to create stakeout information.

Example of Points Created with a Parcel

Creating a point with a parcel is useful when you have a parcel design and need to create stakeout information. In the following illustration, points are created by selecting the line segments of a parcel. The points are placed at the endpoints of each line.

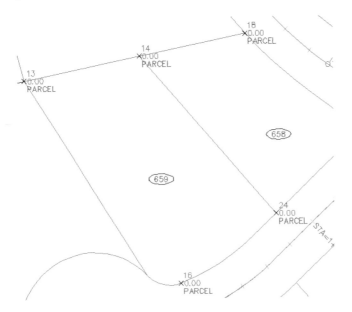

Point Data Created Using Alignments

In addition to parcels and surfaces, another common source of point data is alignments. You can create points that are based on an alignment in several ways, including points that:

- Are offset from alignment stations.

- Are a specified distance apart or equally spaced along an alignment.

- Represent the geometry points on the alignment.

When you create points with an alignment, the application creates points at every geometry point on an alignment, including:

- Points of curvature (PC)
- Points of tangency (PT)
- Spiral curves (SC)
- Curve spirals (CS)
- Tangent spirals (TS)
- Points of intersection (PI)

The raw and full description for the point is automatically assigned based on the type of geometry point that is the source of the new point.

For more information, see "Creating Points Based on Horizontal Alignments" in Help.

Example of Points Created with an Alignment

In this section of an alignment, the points have the following automatically created raw descriptions.

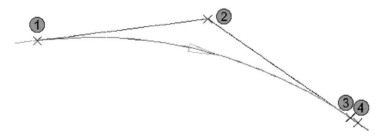

Section of an alignment

1️⃣ Point of curvature

2️⃣ Point of intersection

3️⃣ Point of tangency

4️⃣ Point of curvature

Exercise 01 | Import Points from a Text File

In this exercise, you import points from an external text file to the drawing. The point data was collected by survey crews and provided in a reduced coordinate text file.

The following illustration shows some of the points created for this exercise.

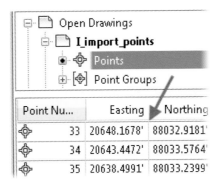

The completed exercise

1   Open \Points\I_import_points.dwg (M_import_points.dwg).

2   Using Windows Explorer, browse to the Points folder.

3   Open I_topopoints.txt (M_topopoints.txt) using a text editor.

   Note the format of the data in the text file. The values represent point number, northing, easting, elevation, and description separated by commas.

4   Close the text editor.

   Next, you import the points to the drawing.

5   On the ribbon, Home tab, click Points > Point Creation Tools.

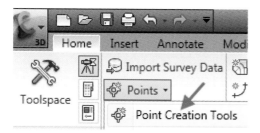

The Create Points toolbar opens. This toolbar contains commands that enable you to create points using a number of different methods.

6   On the Create Points toolbar:

   • Click each down arrow to review the commands available for creating points.

   • Click Import Points.

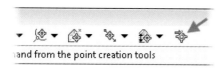

and from the point creation tools

7   In the Import Points dialog box:

   • For Format, select PNEZD (Comma Delimited) from the list.

   • Click the plus (+) sign. Browse to the Points folder.

   • In the Select Source File dialog box, change Files of Type to *.txt.

   • Select *I_topopoints.txt* (*M_topopoints.txt*).

   • Click Open.

8   In the Import Points dialog box:

   • Select the Add Points to Point Group check box.

   • Beside the down arrow, click Add Points to Point Group.

   • In the Point File Formats - Create Group dialog box, for name, enter **Reduced Topo Data - Feb 1 2009**.

   • Click OK twice to close the dialog boxes. Point objects are created in the drawing.

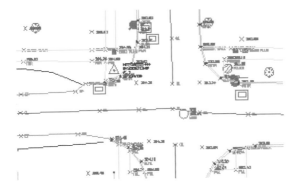

9   Close the Create Points toolbar.

10 In Toolspace, Prospector tab:

   • Click Points.

   • Notice the point data displayed in the item view area.

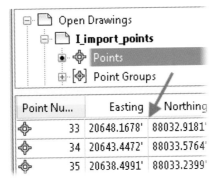

11 In the item view area, right-click any point. Click Zoom To. Civil 3D zooms to the selected
    point(s) in the drawing area.

12 In Prospector:

   • Expand Point Groups. The exclamation marks indicate that the Point Groups are out of
     date and require updating.

- Right-click Point Groups. Click Update. The Point Groups are updated.

13   Click any point group. Notice the points for the point group displayed in the item view area.

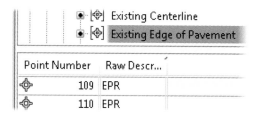

14   Close the drawing. Do not save the changes.

Lesson 12 | Managing Points

This lesson describes how you manage points using description keys, point groups, and point tables. Points are used in different phases of engineering projects and can be difficult to organize and manage. Surveyors and designers sometimes work with thousands of points that have many different descriptions. Description keys and point groups are powerful tools that can help manage your points. Description keys help automate point layering and symbol insertion, while point groups enable you to collectively manage points with similar characteristics.

Objectives

After completing this lesson, you will be able to:

- Describe how point groups are used to identify characteristics shared by points.

- Explain the function of point tables.

- Describe how description keys organize imported data.

- List the guidelines for managing points.

- Create a description key.

- Create points manually.

- Create point groups.

About Point Groups

Point groups provide a flexible and convenient way to identify points that share common characteristics or that are used to perform a task, such as creating a surface. Point groups also play a fundamental role in controlling how a point is displayed in a drawing.

The following illustration shows the Point Group dialog box.

Definition of Point Groups

Point groups organize the points in a drawing and control how they are displayed. You create point groups by defining the characteristics that a point must possess in order to be part of the group. For example, you can specify that a point belongs to a point group based on its point number, its name, its raw or full description, its elevation, or any combination of these parameters. Any point that matches the properties you define is drawn using the point style and point label style for the group.

After you create point groups, you can start to work with points as a group. All actions applied to the group, such as changing the style or assigning the group to a surface, are applied to the individual points in the group. You can export points in a point group to an external text file, or you can delete points within a point group.

Examples for Using Point Groups

Existing Ground Surface Model

You can use point groups to model an existing ground surface. A point group for surface modeling would exclude points that do not represent the terrain, such as manhole inverts and tops of fire hydrants.

Survey Control

A point group for survey control is useful to see which points make up the survey control network.

About Point Tables

Points can contain critical information that sometimes needs to be displayed in a concise format in the drawing area. You can use point table objects to display point data in a table. Point tables are often used for showing survey control points or proposed tree planting schedules.

Definition of Point Tables

A point table is a Civil 3D object that can be used to display point data. The data in the table is organized into columns and is usually sorted by point number. When you insert a point table into a drawing, specified point information is automatically displayed in the table. To show points in the table, you can select point groups or you can graphically select points in the drawing area. The display of the point table is controlled with a point table style.

Point Table Example

It is sometimes useful to show the point data for survey control in a point table in the drawing area as shown in the following illustration. This makes it easy for construction staff to identify monuments and traverse hubs for construction staking.

Survey Control				
Point #	Description	Elevation	Northing	Easting
1	MONV4118UTM	264.12	17888157.44	1820637.55
2	MONV854	261.32	17888139.14	1821307.78
22	LP	262.97	17888098.28	1820644.18
23	LP	263.16	17888102.48	1820644.27
94	LP	264.44	17888157.18	1620645.84
155	LP	269.09	17888331.18	1820650.15

About Description Keys

Description keys help you simplify and standardize the creation of point data in your drawings. Description keys automate the base plan creation process.

For example, when you create points, you can use a description key to assign the point style, a point label style, or to place the points on a specific layer. Description keys help you simplify and standardize the creation of point data in your drawings. After you create description keys using your organization's standards, you can save them as a part of a drawing template (.dwt). When you use the template as the starting point for your drawings, the description keys are carried forward to the new drawing.

The following illustration shows how description keys manage imported points.

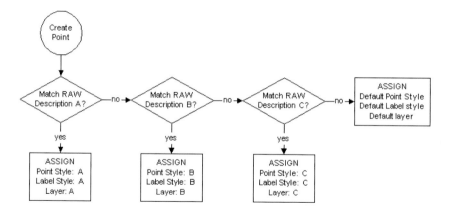

Definition of Description Keys

Description keys are lookup tables that you use to assign the following:

- **Alternate Point Descriptions**

Surveyors collect data with field descriptions, also known as raw descriptions. Field descriptions are usually abbreviated and can sometimes be numeric. When you assign an alternate point description, you can expand on the abbreviated field description to make the description more meaningful to those in the office.

- **Point Style**

Controls the display of the point node, or marker. The point style can reference an AutoCAD® block and can be used to display base plan symbols such as catch basins, trees, and manholes.

- **Point Label Style**

You can label points with elevation only, description only, point number only, or any combination of these and other parameters with point label styles.

- **Layer**

The layer the point is created on. Description keys are organized in description key sets and can be saved in the drawing template.

Example of Description Keys

An organization can create a single set of description keys and use this set to import data from a number of organizations. For example, a description key set can have entries for the raw descriptions APPLE* and TREE*. These entries ensure that an appropriate point style and other properties are assigned whether the source data has a category for each specific type of tree or a single generic tree category. Instead of creating a description key set for each source of data, a firm can use the same set of description keys any time point data is imported.

Process: Creating and Editing Description Keys

The following steps describe how to create and edit description keys.

1   On the Settings tab, click to expand Point. Right-click Description Key Sets and click New for creating a new description key, or right-click the name of an existing set and click Edit Keys to edit the existing set.

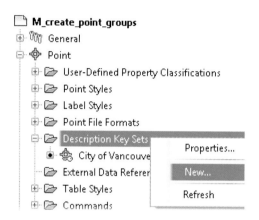

2   Enter a name for the new set. Enter a brief description.

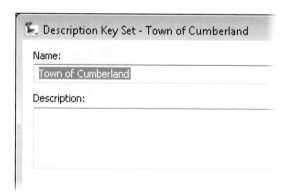

3   In the Description Key Editor, right-click a description key in the Code column and click New. Click anywhere in the row to select the description key you want to edit. Click each field along the row to edit values.

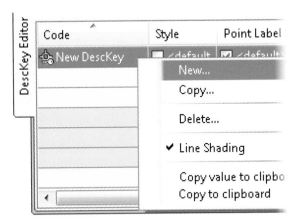

Guidelines for Managing Points

Keep the following guidelines in mind when you use description keys, point groups, and point tables to manage points.

Description Keys Guideline

- You should take into account your organization's standards when you create description keys.

Point Groups Guideline

- If you have points that share common display characteristics, use a point group to identify the point style and point label style for these points, instead of assigning a point style and a point label style to each individual point.

Exercise 01 | Create a Description Key

In this exercise, you create a catch basin description key in an existing description key set.

The completed exercise

1   Open *\Points\I_description_keys.dwg* (*M_description_keys.dwg*).

2   In Toolspace, on the Settings tab, click to expand Point. Click to expand Description Key Sets.

    Notice the existing description key set named City of Vancouver.

3   Right-click City of Vancouver. Click Edit Keys.

    The Panorama window shows the DescKey Editor with the existing description keys.

4   Right-click any column header. Clear every column except Code, Style, Point Label Style, Format, and Layer.

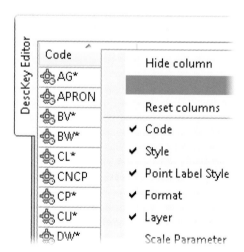

5   Double-click the headers at the column lines to expand the columns for better viewing.

6 Right-click any description key. Click New.

A new description key entry, New Desckey, appears in Panorama, and is located alphabetically.

7 Enter the following parameters for the new description key:

- For Code, enter **CB**. Use uppercase. (**Note**: The description keys re-sort their order).

- Select Style to enable the option.

- Click in the Style cell to the right. In the Point Style dialog box, select STM CB from the list. Click OK.

- In the Style cell, select the check box. Click in the style cell to pick the style. In the Point Style dialog box, select STM CB from the list. Click OK.

- For Format, enter **CATCH BASIN**.

- Press ENTER.

8 Click the green check mark to close the Panorama window and save the changes.

9 Close the drawing. Do not save the changes.

Exercise 02 | Create Points Manually

In this exercise, you add catch basin points on the curb returns of an existing intersection. You may need to add points manually when survey data was incomplete. The location of manually created points is often identified by a position relative to two other points, or a position relative to an intersection curb return.

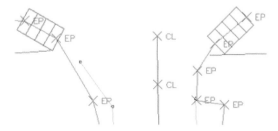

The completed exercise

1   Open \Points\I_create_points_manually.dwg (M_create_points_manually.dwg).

2   In Toolspace, on the Prospector tab, click Points.

3   In the item view area, click the Point Number column header a few times to sort the points by point number.

4   To view some key areas on the drawing:

   • Scroll down to point 732.

   • Right-click 732. Click Zoom To.

   Notice points 741 and 742 on the left side of the intersection. Also notice points 606 and 607 on the right side of the intersection.

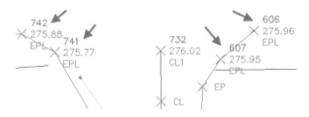

Catch basin points are required on the curb returns midway between these points.

5   On the ribbon, Home tab, click Points > Point Creation Tools.

6   On the Create Points toolbar, click Manual.

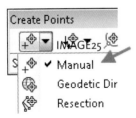

7   With your cursor in the drawing area, press SHIFT+right-click. Click Midpoint.

8   To create the first catch basin:

  • Hold the mouse over the polyline running between points 741 and 742.

  • When you see the triangular midpoint object snap marker, select the polyline.

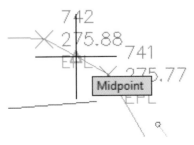

  • When prompted for a point description, enter **CB**.

  • Press ENTER.

  • When prompted to specify a point elevation, enter **275.83' (84.07 m)**.

  • Press ENTER.

9   Repeat the steps above to add a catch basin between points 606 and 607.
    Use **275.9' (84.05 m)** for the elevation.

10 Press ENTER to end the command.

    Two catch basin points are created. Notice that the point style shows the catch basin symbol
    and the point label style shows just the description.

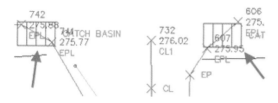

11 In the drawing area:

    • Click points 732, 741, 742, 606, and 607 so that they are all selected.

    • Right-click. Click Edit Points.

12 The five points are displayed in the Panorama window. To edit the point label style:

    • Using the SHIFT key, select the five points.

    • Scroll to the right.

    • Right-click the Point Label Style column header.

    • Click Edit.

    • In the Select Label Style dialog box, select _No Display.

    • Click OK.

    The point label is turned off.

13 Close the Panorama window.

14 To adjust the rotation of the catch basin points:

   - In the drawing area, click a catch basin point.

   - Click the green circular grip.

   - Rotate the point so the symbol aligns with the pavement edge.

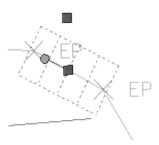

   - Click again to release the point.

15 Close the drawing. Do not save the changes.

Exercise 03 | Create Point Groups

In this exercise, you organize point data by creating point groups. Site development projects usually involve thousands of points. These points are used for a number civil engineering tasks including survey data reduction, base plan creation, surface modeling, design, and construction. Point data management becomes a task itself.

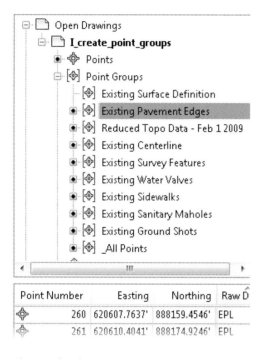

The completed exercise

1 Open file *Points\I_create_point_groups.dwg* (*M_create_point_groups.dwg*).

2   In Toolspace, on the Prospector tab:

   • Click to expand Point Groups.

   Notice the _All Points point group. This point group always exists, is not editable, and contains all of the points in the drawing.

3   Click _All Points.

4   In the item view area, adjust the columns so that you see Point Number, Northing, Easting, Raw Description, and Full Description.

5   Click the Raw Description column header to sort the list by Raw Description.

6   Scroll down to the edge of pavement shots.

   The edge of pavement shots have a raw description of EPL and EPR.

	765	620612.6639'	888794.6129'	EPL
	766	620618.1616'	888794.8720'	EPL
	767	620618.5963'	888797.2095'	EPL
	109	620631.8914'	888161.8647'	EPR
	110	620632.1567'	888167.9910'	EPR
	111	620632.1845'	888170.5936'	EPR

7   Next, you create a point group for the existing pavement edges that includes points with the EPR and EPL descriptions.

   • Right-click Point Groups. Click New.

8   In the Point Group Properties dialog box, Information tab:

- For Name, enter **Existing Pavement Edges**.

- Click Apply.

9   On the Raw Desc Matching tab, for Code, select EP*. A check mark is displayed next to the option.

The Raw Desc Matching tab shows entries in the description key set. It is easier to create point groups when a description key set exists.

10   On the Include tab, notice that With Raw Descriptions Matching is toggled on and that there is an EP* entry.

☐ With names matching: [   ]

☑ With raw descriptions matching   EP*

☐ With full descriptions matching: [   ]

Alternatively, you could go directly to the Include tab. Select With Raw Descriptions Matching. Enter **EP***.

The asterisk (*) is a wild card indicating that all descriptions with an EP prefix are included. Alternatively, you could enter **EPR, EPL**.

11   Click the Point List tab.

Note that the points with descriptions EPL and EPR are included in the point group.

- Click OK.

In Prospector, a new point group is created and appears under Point Groups.

- [⊕] Point Groups
  - [⊕] Existing Pavement Edges
  - [⊕] Reduced Topo Data - Feb 1 2009
  - [⊕] Existing Water Features
  - [⊕] Existing Sidewalks
  - [⊕] Existing Survey Features
  - [⊕] Existing Centerline
  - [⊕] _All Points

12 Right-click Existing Pavement Edges. Click Properties.

13 Explore the Point Group Properties dialog box. Click Cancel.

14 Right-click Existing Pavement Edges. Note the other menu options.

Next, you create a point group for surface modeling. For this point group, an exclusion criterion is used to exclude certain points from the surface model. These points are the monuments, lead plugs, and fire hydrants.

The elevations of these points are not suitable for surface modeling because they do not adequately represent the existing terrain elevation.

15 Right-click Point Groups. Click New.

16 In the Point Group Properties dialog box, Information tab, for Name, enter **EG Surface**.

17 On the Exclude tab:

- Select the With Raw Descriptions Matching option.

- Enter **MON*, LP, FH***.

☑ With raw descriptions matching: MON*, LP, FH*

☐ With full descriptions matching:

18   On the Point List tab, note that these points are not in the list. Click OK.

19   Close the drawing. Do not save the changes.

Chapter 04
Surfaces

This chapter describes how you work with surfaces. Surfaces are used to represent existing surface and sub-surface terrain, and proposed design surfaces.

The Creating Surfaces lesson describes how you create a surface for an urban road using point group data and breakline data. The Modifying Surfaces lesson describes how you modify the properties of a surface to change the surface style, manage the crossing breaklines and change the surface definition parameters. You also edit a surface to delete the long and inaccurate triangulation lines that span across the corners of the intersections. Finally, the Creating Surface Styles lesson describes how you create and apply a surface style that shows contours at major and minor increments.

▶ **Objectives**

After completing this chapter, you will be able to:

- Create surfaces and use breaklines to add detail to surfaces.

- Modify surfaces by editing the surface properties and surface TIN lines.

- Create, modify, and apply surface styles.

Lesson 13 | Creating Surfaces

This lesson describes how to create surfaces from points, and how to use breaklines to add detail to surfaces to define the terrain breaks.

You create surfaces to represent both existing and proposed conditions. You can create surfaces from different data sources including point groups, points in an external file, breaklines, contours, and AutoCAD® objects. The most common way to select points for a surface is through a point group.

Provided that proper field procedures and standards are in place, surveyors can create surfaces as a direct by-product of their field observations.

The following illustration shows a surface.

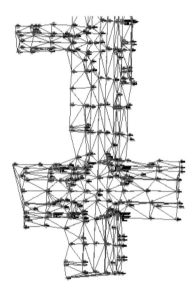

Objectives

After completing this lesson, you will be able to:

- Describe surfaces.

- Describe how breaklines are used for surface modeling.

- List the guidelines for surface modeling.

- Create a surface from points and breaklines.

About Surfaces

Surfaces are the basic building blocks in Civil 3D®. You use surface objects to create a three-dimensional representation of existing and proposed project data.

A Triangulated Irregular Network, or TIN, is a surface model consisting of data points (vertices) connected by 3D-lines (TIN lines) to form three-dimensional irregularly shaped triangular faces. These triangular faces are collectively called a TIN. TINs are used to model existing ground surfaces, proposed surfaces, subsurfaces (like bedrock), and water surfaces.

Definition of Surfaces

Surfaces are intelligent three-dimensional representations of either existing or proposed terrain. By connecting surface points with a network of triangulation lines, you create a dynamic, continuous representation of a surface. Using this network, you can interpolate the elevation of any location on the surface, not just locations that are defined by a point. The TIN models a continuous surface that you can use to display contours and elevation data, and create surface profiles and targets for slope daylighting.

You can create two types of surfaces. Regular surfaces are used to model existing and proposed terrain. You use volume surfaces to calculate site volumes and reference an existing and a proposed surface.

The following illustration shows a TIN volume surface being created.

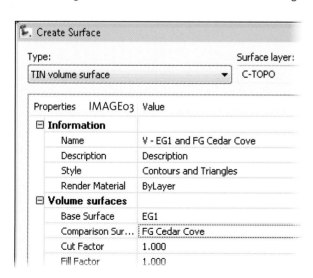

Examples of Surfaces

You use surfaces that represent existing conditions to:

- Display base mapping contours.
- Generate surface profiles for alignments.
- Generate surface sections.
- Calculate volumes when compared against a proposed surface.
- Calculate grading limits for corridors and grading objects.

You use surfaces that represent proposed conditions to:

- Display proposed contours.
- Calculate volumes when compared against an existing surface.
- Generate construction staking data.
- Label design spot elevations and grades.

The following illustration shows a surface that represents an existing road intersection.

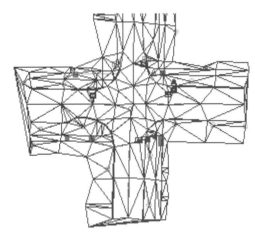

About Breaklines

Breaklines are required for most surface modeling tasks. Surface triangulation lines follow along the defined breaklines. It is important that triangulation lines follow the terrain breaks in order to accurately model the terrain.

When you define a breakline, the surface lines follow the breakline instead of interpolating an elevation for the location using the closest points. Breaklines increase the accuracy of the surface because they can always interpolate elevations between two consecutive points along a distinctive terrain break. Adding a breakline improves the three-dimensional representation of linear features on the surface.

Definition of Breaklines

Breaklines are used to define distinctive breaks in the terrain such as road crowns, ditch bottoms, gutter lines, and banks. The two most common types of breaklines are proximity breaklines and standard breaklines.

Proximity Breaklines

Proximity breaklines are defined from 2D polylines. Elevations and locations of breakline vertices are determined by physically relocating each vertex of the 2D polyline to the nearest point used to create the surface, and assigning the northing, easting, and elevation of that point to the vertex.

Proximity breaklines are most often used when the polylines are created by snapping to the nodes of the points. This functionality is very useful when you want to use 2D polyline base plan entities (pavement edge, crown, ditch bottom, and so on) as breaklines.

Standard Breaklines

Standard breaklines are defined from 3D polylines. The northing, easting, and elevations of the polyline vertices are used to define the points along the breakline. Breaklines are listed on the Prospector tab of the Toolspace windows by expanding Surface, Surface Name, Definition, and Breaklines.

Example of Using a Breakline to Delete Triangles

Suppose that your surface data includes a line of points representing a ridge or swale, but the points are far apart in places. As a result, there might be triangles that cross the line, reducing the precision of the surface. When this happens, you can create a breakline that represents the feature more accurately.

In another example, you are designing a surface feature that represents a pond. You can place a breakline to stop triangulation at the edge of the pond, or you can eliminate triangles within the pond.

In the following illustration, the 3D polyline represents the course of a stream. Triangles cross the stream at various points.

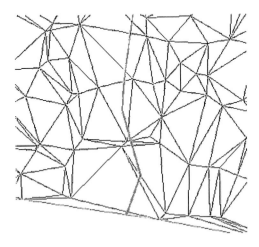

Surface with 3D polyline

When you use the 3D polyline to create a standard breakline, the elevations at each of the line's vertices are added to the surface. When you rebuild the surface, no triangles cross the breakline.

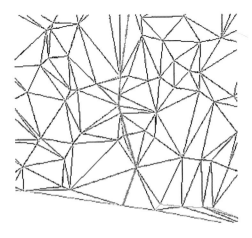

Surface with breakline

Chapter 04 | Surfaces

Notice how the polyline crosses the triangles that describe the surface in the illustration on the left. The illustration on the right shows the same polyline made into a breakline. Notice how the surface triangles follow the breakline, increasing the accuracy of elevation calculations at this part of the surface.

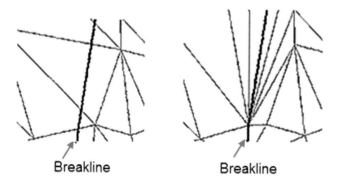

Guidelines for Creating Surfaces

The following guidelines can help increase your productivity when you create surfaces.

- When creating surfaces from point data, points that do not represent the topography should be excluded from the surface model to ensure the accuracy of the surface. Examples of points to exclude include manhole inverts, hydrant tops, and raised manhole lids. You can use the exclusion criteria in a point group to exclude certain points from the surface model.

- If the surface contains terrain breaks, breaklines must be defined to accurately represent the surface.

Exercise 01 | Create a Surface

In this exercise, you create a surface from points and breaklines. A point group for surface modeling already exists in the drawing. This point group excludes the points that are not suitable for surface modeling such as tops of fire hydrants and survey monument data. Breaklines are created using the proximity breakline definition method.

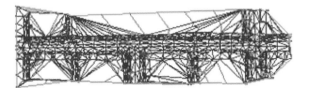

The completed exercise

1   Open \Surfaces\I_surfaces.dwg (M_surfaces.dwg).

2   In Toolspace, on the Prospector tab:

   • Click to expand Point Groups.

   • Right-click Existing Surface Definition. Click Properties.

3   In the Point Group Properties dialog box, on the Exclude tab:

   Notice the exclusions. The points matching these descriptions do not represent the elevation of the existing topography and are excluded from the surface. (Note: The asterisk '*' is a wild card).

   • Click Cancel.

   • First, you create the surface.

4   In Prospector:

   • Collapse Point Groups.

   • Right-click Surfaces. Click Create Surface.

5   In the Create Surface dialog box, under Information:

   • For Name, enter **EG1**.

   • For Description, enter **Existing Ground Topo**.

- Note that the default Style is Contours and Triangles. The assignment of the default surface style is done in Command Settings.

- Click OK.

In Prospector, EG1 is added to the Surfaces collection.

6   In Toolspace:

- Click to expand Surfaces, EG1, and Definition.

- Right-click Point Groups. Click Add.

7   In the Point Groups dialog box:

- Click Existing Surface Definition.

- Click OK.

The surface is built and displayed using the Contours and Triangles surface style.

Notice the points that were excluded from the surface model.

8   Zoom in to the four-legged intersection on the right side. Notice the points that were excluded from the surface model.

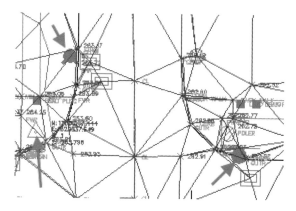

9   Look closely at the triangulation and the base plan linework. All the polylines on the base plan represent terrain breaks. The triangulation lines, however, do not follow over the base plan polylines.

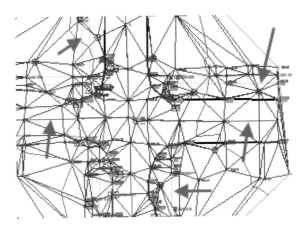

10   Navigate to other areas in the drawing and make the same observation.

Next, you define breaklines to force the triangulation along the terrain breaks.

11   In Prospector, under Surfaces, EG1, Definition, right-click Breaklines. Click Add.

12   In the Add Breaklines dialog box:

   - For Description, enter **All Breaklines**.

   - For Type, click Proximity.

   - Click OK.

Proximity breaklines are defined from 2D polylines. When defining proximity breaklines, elevations are assigned to the polyline vertices by relocating each polyline vertex to the nearest DTM point, and the point elevation is assigned to the vertex. Most vertices are drawn to the points, so vertices do not move but the point elevation is assigned.

13   When prompted to select objects, use the AutoCAD window selection set option and select the entire drawing to select all 2D polylines. Press ENTER.

14 If an exclamation mark appears next to EG1, right-click EG1. Click Rebuild.

15 The EG1 surface automatically updates. Notice that the triangulation lines follow directly over the base plan polylines. This indicates that all breaklines are defined.

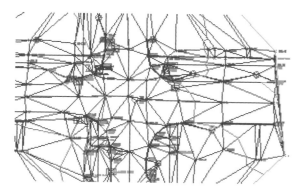

16 You receive some warning and error messages in Panorama. Review the warning and error messages.

The error messages indicate crossing breaklines. This occurs in the locations where the breaklines are slightly overlapped. Do the following:

- Click Action > Clear All Events.

- Close Panorama.

17 Close the drawing. Do not save the changes.

Lesson 14 | Modifying Surfaces

This lesson describes how you modify surfaces by editing the surface properties and surface TIN lines. You modify Surface Properties to assign a surface style, change the surface definition parameters, specify surface analysis parameters, and to review surface statistics. You can also perform other useful functions such as setting the maximum triangle length and resolving crossing breaklines. You edit the surface TIN lines to ensure that the surface correctly represents the topography.

Objectives

After completing this lesson, you will be able to:

- List the methods for modifying surfaces.

- Edit surface triangles.

- List the guidelines for modifying surfaces.

- Modify surface properties.

- Edit a surface to ensure that the triangulation lines correctly represent the topography you are modeling.

Methods for Modifying Surfaces

There are two methods for modifying surfaces:

- Modify surface properties.

- Modify surface triangles.

Surface Properties

You can make changes to a surface by editing items in the Surface Properties dialog box:

- Provide name and description.

- Assign a surface style.

- Lock or unlock a surface.

- Specify the maximum triangle length for a surface.

- Resolve crossing breaklines.

- Specify analysis parameters.

- Review surface statistics.

Surface Triangles

To modify a surface, you can also edit the triangles that define the surface. The types of changes you can perform include deleting the triangles and swapping the triangle edges.

Operation Type List

The Operation Type list displays all of the data addition and editing operations made to a surface, in the order in which they were performed. It also shows the parameters for each operation.

When you create and edit a surface, you can perform many different types of operations to add data and edit surfaces. The order and inclusion of these operations could produce undesired results. The Operation Type list makes it easy to track the operations you perform on a surface, undo or redo the operations, or change the order in which operations are executed.

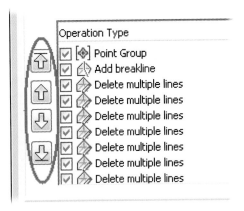

Maximum Triangle Length

When you set the maximum triangle length, you can eliminate long triangulation lines near the outer border of the surface.

This first illustration shows a surface without the maximum triangle length set.

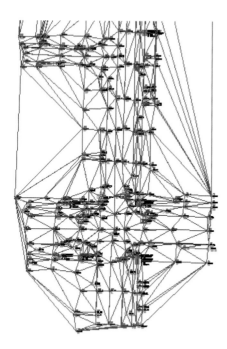

This second illustration shows the surface with the maximum triangle length set to 75' (20m).

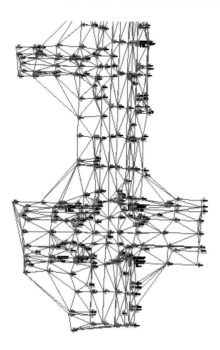

Crossing Breaklines

By definition, triangulation lines are forced along breaklines. Furthermore, triangulation lines cannot cross each other. Therefore, when breaklines cross an error is generated. If a resolution to the breakline is not assigned, the second breakline created will be ignored.

You can resolve crossing breaklines using any of the following methods:

- Use the first breakline elevation at the intersection.

- Use the last breakline elevation at the intersection.

- Use the average breakline elevation at the intersection.

The most common solution for crossing breaklines is to enable the Use Average Breakline Elevation at the Intersection option. You enable this option in the Surface Properties dialog box, shown in the following illustration.

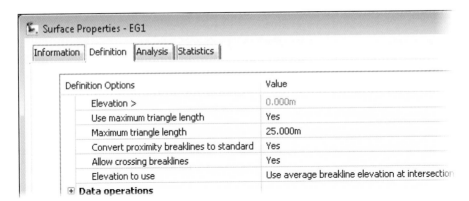

Editing Surface Triangles

You can edit surface triangles to achieve greater precision in your terrain model. You can delete triangles or change the orientation of their edges. An edit operation list is maintained so that you can view different versions of the surface as you edit.

Deleting Triangles

In some cases, triangles are created in areas where there is not enough point data to accurately represent the surface shape and elevation. As part of your review of the surface data, you can identify the superfluous triangles and delete them from the surface.

The following illustration shows a section of a surface that was created by importing point data from a text file without defining a boundary. The long triangulation lines located at the surface boundary are created by joining two distant points. The point data in this area is too scarce to include it in the surface. To refine the surface, you can create a boundary that excludes the problem area, or you can delete the triangles.

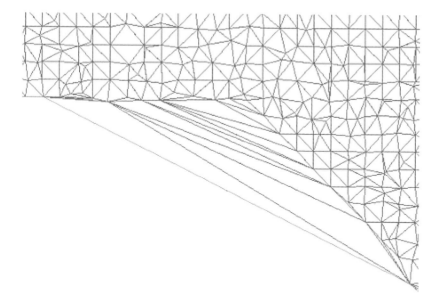

The following illustration shows the same area of the surface after you delete the triangles. The surface border now coincides with the limit of the point data.

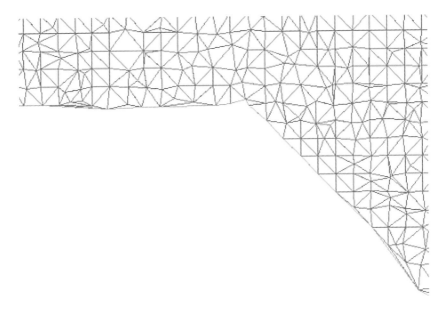

Swapping Triangle Edges

Often you are provided with contour or polyline data to create a surface. After you import the contours, a surface is created that consists of a network of triangles based on the X, Y, Z coordinates of the contour vertices. Sometimes, the surface triangles are oriented incorrectly, producing a flat surface rather than a three-dimensional one. You can configure Civil 3D to automatically check for these flat locations. Alternatively, you can change the direction of the triangle lines using the Swap Edge command.

The following illustration shows a detail view of a surface created from contours. In the circled area, the generated triangles connect points on a contour, rather than connecting points between contours. This location on the surface appears as though there was no change in the elevation, showing a flat surface.

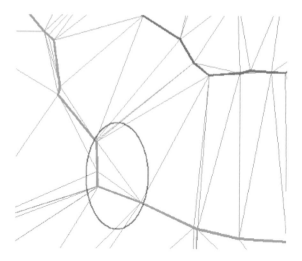

With the Swap Edge command, you can change the direction of triangle lines to improve the accuracy of the surface. In the following illustration, two triangle edges that connected the same contour were edited. In the original version of the surface, the contours were oriented from upper left to lower right. In the current version, they are oriented from lower left to upper right.

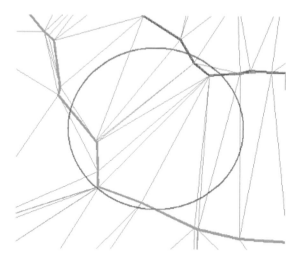

For more information about swapping edges, see "Edge Swapping" in Help.

> *You cannot swap an edge that is part of a breakline or an edge that you create using the Add Line command.*

Guidelines for Modifying Surfaces

Keep the following guidelines in mind when you modify surfaces.

- To keep surface editing to a minimum, create a surface from quality data. Always use breaklines and remove data that does not accurately represent the terrain.

- When you edit a surface, it is useful to assign a style that displays both triangles and contours. This style will enable you to see the effects of editing triangulation lines on the contours.

- Best practices in breakline creation are breaklines that do not cross. In some instances, however, you may be provided with breaklines from another source that may cross each other in some locations. In these cases, it is necessary to resolve the crossing breaklines so that you create an accurate surface.

Exercise 01 | Modify Surface Properties

In this exercise, you modify the properties of the surface to:

- Change the surface style.

- Manage crossing breaklines.

- Change surface definition parameters.

- Review surface statistics.

The completed exercise

1   Open \Surfaces\I_surface_properties.dwg (M_surface_properties.dwg).

2   In Toolspace, on the Prospector tab:

- Click to expand Surfaces.

- Right-click EG1. Click Surface Properties.

3   In the Surface Properties - EG1 dialog box, on the Information tab:

- For Surface Style, select Contours 1' and 5' Background (Metric: Contours .2m and 1m Background).

- Click OK.

The display of the surface updates to show contours.

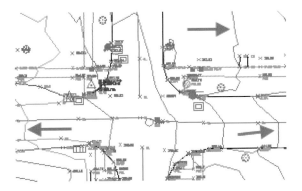

4   In the drawing area:

 •   Click any contour to select the surface.

 •   Right-click. Click Surface Properties.

5   In the Surface Properties - EG1 dialog box, on the Information tab:

 •   For Surface Style, select _No Display.

 •   Click OK.

The drawing updates and the display of the surface is suppressed.

6   In Prospector:

 •   Click to expand Surfaces.

 •   Right-click EG1. Click Surface Properties.

7   In the Surface Properties - EG1 dialog box, on the Information tab:

 •   For Surface Style, select Elevation Banding (2D).

 •   Click Apply.

8   Drag the Surface Properties dialog box away. Notice the changes in the drawing.

9   In the Surface Properties dialog box:

- For Surface Style, select Contours and Triangles.

- Click Apply.

Note the changes in the drawing.

Next, you set the maximum triangle length.

10   Click the Definition tab. Click to expand Build and do the following:

- For Use Maximum Triangle Length, select Yes.

- For Maximum Triangle Length, enter **75' (30 m)**.

- For Allow Crossing Breaklines, select Yes.

- For Elevation to Use, select Use Average Breaklines Elevation at Intersection. This action eliminates the crossing breaklines processing error that you encountered earlier.

- Click OK.

Definition Options	Value
Elevation >	0.000'
Use maximum triangle length	Yes
Maximum triangle length	75.000'
Convert proximity breaklines to standard	Yes
Allow crossing breaklines	Yes
Elevation to use	Use avera

11   In the Surface Properties - Rebuild Surface dialog box, click Rebuild the Surface.

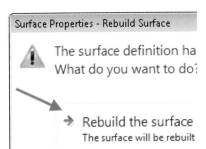

In the drawing, all triangles shorter than 75' (30m) are removed. This technique is useful for deleting long unwanted triangles near the perimeter of the surface.

12   In Prospector, right-click EG1. Click Surface Properties.

13   In the Surface Properties - EG1 dialog box, on the Statistics tab:

- Click to expand General.

- Note the elevation statistics for the surface.

- Click to expand Extended and TIN.

- Review the other surface statistics.

- Click OK.

Your drawing at this stage appears as follows:

14   Close the drawing. Do not save the changes.

Exercise 02 | Edit a Surface

In this exercise, you edit a surface to ensure that the triangulation lines correctly represent the topography you are modeling. You add a line, delete a line, and swap an edge on a surface.

The completed exercise

1   Open \Surfaces\I_edit_surface.dwg (M_edit_surface.dwg).

2   Zoom in to the intersection on the right side of the drawing.

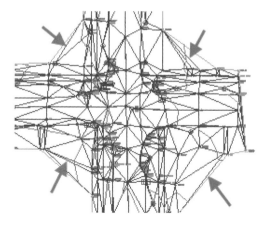

Notice that in the vicinity of the intersections, some longer triangles remain and span across corners. These triangles are not accurate and result in the incorrect display of contours.

3   In Toolspace, on the Prospector tab:

   • Click to expand Surfaces, EG1, Definition.

   • Right-click Edits. Click Delete Line.

4   On the command line, enter **C**. Press ENTER.

5   Using the crossing selection method (click and drag the mouse to the left), select some of the longer unwanted triangulation lines that span across the intersection corners. Press ENTER when done.

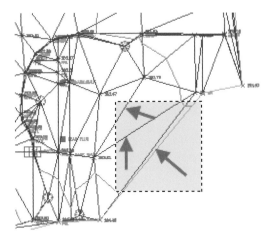

Tip: You can use any AutoCAD selection set option such fence or a single object selection.

6   Notice how the contours automatically update to reflect the surface changes.

7   In Prospector, right-click Edits. Click Swap Edge.

8   When prompted to select an edge, select a triangulation line with a contour passing through it. Press ENTER when done.

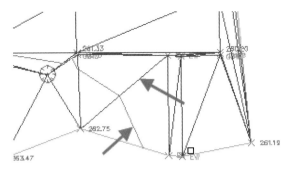

9   Notice how the edge reorients itself in the opposite direction. If there is a contour present, it also readjusts based on the position of the triangle.

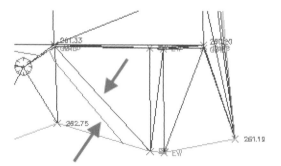

10   In Prospector, click Edits.

Notice the list of edits in the item view area.

Drawing Objects	
● 🔷 Edits	

Edits	Description
🔷 Delete multiple lines	First passes through (16206
🔷 Delete multiple lines	First passes through (16206
🔷 Swap Edge	At (1620592.7083',17888057
🔷 Swap Edge	At (1620589.7491',17888057
🔷 Swap Edge	At (1620592.7083',17888057

11   Right-click any of the edits. Notice that you can delete any unwanted edits.

Note: If you do not have the option to delete an edit in the item view, you can make the change in the Surface Properties, Definition tab. Right-click the operation, and select Remove from Definition.

12   If you have time, use these two commands to finish editing the EG1 surface. Your final drawing should appear as follows:

13   Close the drawing. Do not save the changes.

Lesson 15 | Creating Surface Styles

This lesson describes surface styles and how to modify, create, and apply them.

You create and assign surface styles based on how you want to display the surface. For example, if you need to edit a surface, you assign a surface style that shows both triangulation lines and contours. If you want to display a surface as part of a background map, you assign a surface style that displays just contours. You can even assign a surface style that suppresses the display of the surface entirely.

Objectives

After completing this lesson, you will be able to:

- Describe surface styles and how they affect the model display.

- List the guidelines for creating surface styles.

- Create surface styles.

About Surface Styles

You assign a surface style to a surface to control how the surface object is displayed. There are many uses for surface styles. You can use surface styles to help you edit a surface, visualize a surface, analyze a surface, or display a surface on a final set of contract plans. With model-based design, design objects can serve as the final drafted objects that you would display on contract drawings.

The following illustration shows a surface that uses a surface style to display contours.

Definition of Surface Styles

Surface style controls the display of a surface object. You select a surface style that make it easier to edit, visualize, analyze, or display a surface on a final set of contract plans.

Surface Display Styles

You can create surface styles to display any combination of the following surface components:

- Points
- Triangles
- Borders
- Contours (major, minor, and user)
- Grids
- Analysis (directions, elevations, slopes, slope arrows, and watersheds)

You can also create surface styles that suppress the display of all surface components. This would be named as a No Display style and is an alternative to turning off the surface layer.

Surface Style Example: Contours and Triangles

A surface style that displays contours and triangles is useful for editing a surface. When you edit a surface, you manipulate the triangulation lines. If the contours are also visible, you can immediately see the effects of the edits on the contours.

The following illustration shows the Surface Style dialog box for a style that displays the Triangle, Major Contour, and Minor Contour components.

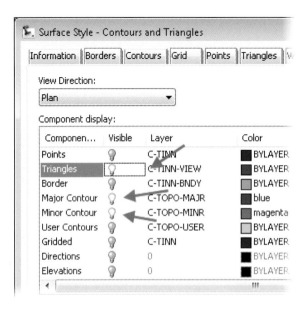

A surface using the Contours and Triangles surface style is shown in the following illustration.

Guidelines for Creating Surface Styles

Keep the following guidelines in mind when creating surface styles.

- You create surface styles for both design and final plotting tasks. A surface style for designing may show triangles and contours, whereas a surface style for final plotting may show just contours.

- Use the No Display style as an alternative to using layers to suppress the display of a surface. This method is quicker and reduces the need for creating additional layers.

Exercise 01 | Create Surface Styles

In this exercise, you modify, create, and apply surface styles. A surface is a singular object in Civil 3D. The display of a surface is controlled by using a surface style. Surfaces styles can be used to display the surface as points, triangles, borders, contours, and grids.

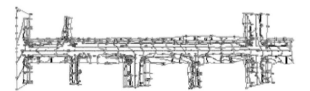

The completed exercise

1 Open \Surfaces\l_surface_styles.dwg (M_surface_styles.dwg).

The surface in the drawing is displayed using the Contours and Triangles surface style. Note that not many contours are visible. The contour interval is too big. You begin by modifying the Contours and Triangles surface style to decrease the contour interval.

2 In Toolspace, on the Settings tab:

- Click to expand Surface and Surface Styles.

Notice the collection of surface styles.

Note the yellow triangle next to the Contours and Triangles surface style. The triangle indicates the style used to display the EG1 surface.

- Right-click Contours and Triangles. Click Edit.

3 In the Surface Style dialog box, click the Display tab and do the following:

- Notice that this surface style displays the Triangles, Border, Major Contours, and Minor Contour surface components.

- For Major Contour, change the color to Blue (color 5).

- For Minor Contour, change the color to Magenta (color 6).

Border	C-TINN-BNDY	BYLAYER
Major Contour	C-TOPO-MAJR	blue
Minor Contour	C-TOPO-MINR	magenta
User Contours	C-TOPO-USER	BYLAYER

4   Click the Contours tab and do the following:

- Expand Contour Intervals.

- For Minor Interval, enter **0.5 (0.1 m)**.

- For Major Interval, enter **2.0 (0.5 m)**.

- Click OK.

The surface display updates to reflect the decreased contour interval and the changed colors.

Next, you create a new surface style that only displays triangles.

5   In Settings, click to expand Surface, Surface Styles. Right-click Surface Styles. Click New.

6   In the Surface Styles – New Surface Style dialog box:

- Click the Information tab.

- For Name, enter **Triangles**.

- Click the Display tab.

- Turn all components off, except Triangles.

- For Triangles, change the color to red.

- Click OK.

Next, you apply the surface style.

7   In the drawing area, click the surface. Right-click and then click Surface Properties.

8   In the Surface Properties dialog box, Information tab, for Surface Style, click Triangles. Click OK. The surface display updates.

Next, you review and apply the _No Display surface style.

9   In Toolspace, on the Settings tab:

- Click to expand Surface, Surface Styles.

- Right-click _No Display. Click Edit.

10 In the Surface Style - _No Display dialog box, on the Display tab:

- Notice that all components are not visible.

The _No Display surface style can be assigned to surfaces to suppress their display.

- Click Cancel. Next, you apply the _No Display surface style to the EG1 surface.

11 In the drawing area, select the surface. Right- click and then click Surface Properties.

12 In the Surface Properties - EG1 dialog box, on the Information tab:

- For Surface style, select _No Display.

- Click OK.

AutoCAD® Civil 3D® suppresses the display of the surface. _No Display styles are a useful alternative to turning off a layer.

Next, you assign a style to display just the contours.

13　In Toolspace, on the Prospector tab:

- Click to expand Surfaces.

- Right-click EG1. Click Surface Properties.

14　In the Surface Properties - EG1 dialog box, on the Information tab:

- For Surface Style, click Contours 1' and 5' Background (Contours 0.2m and 1m (Background).

- Click OK. The surface display updates.

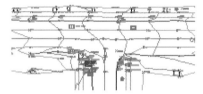

15　Look closely at the contours and notice the sharp corners. Next, you modify the surface style to introduce contour smoothing.

16　In the drawing area, select the surface. Right- click and then click Edit Surface Style.

17　In the Surface Style dialog box, click the Contours tab and do the following:

- Click to expand Contour Smoothing.

- For Smooth Contours, change the value to True.

- At the bottom of the dialog box, using the slider, maximize the contour smoothing factor.

- Click OK.

The display of the surface updates to show smoothed contours.

18　Close the drawing. Do not save the changes.

Chapter 05
Site Design Parcels

This chapter describes how you create and work with parcels for a residential subdivision design.

The Creating Sites lesson describes how you create a site for future parcels. The Creating Right-of-Way Parcels lesson describes how you use the create right of way command to create a road allowance for an alignment. The Creating Parcels lesson describes how you create parcels using the layout tools and from existing objects. The Editing Parcels lesson describes how you edit parcel segment geometry by adding and deleting parcel segments and by renumbering them. Finally, the Labeling Parcel Segments and Creating Tables lesson describes how you label the parcel segments with line and curve geometry details. You also learn how to create tag labels on the parcel segments and produce a related table that displays the parcel data.

▶ **Objectives**

After completing this chapter, you will be able to:

- Create sites with design objects.

- Create right-of-way parcels.

- Create parcels from objects or by using the Parcel Layout Tools.

- Modify parcel objects using editing and renumbering commands.

- Label parcel segments with geometry data or tags, and create parcel data tables.

Lesson 16 | Creating Sites

This lesson describes how to create a site that contains design objects, such as alignments, parcels, and grading groups. Using a site, you can organize boundaries, alignments, and parcels in a drawing. In many cases, this site is also the initial parcel from which all subsequent parcel design is generated.

Objectives

After completing this lesson, you will be able to:

- Explain the function of sites.

- List the guidelines for creating sites.

- Create a site.

About Sites

You create and maintain relationships among objects by grouping them together in a site. The following illustration shows a site and the objects that make up that site.

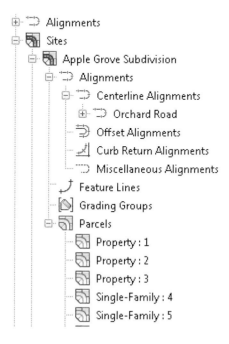

Definition of Sites

A site is a collection of design objects such as alignments, parcels, feature lines, and grading groups. When you create any of these objects, you must specify the site to which it belongs. If you create these objects before you create a site, a site with a default name (for example, Site 1) is created automatically, and the objects are assigned to it.

Sites are most often used to organize data. Sites also recognize topology. Topology refers to the spatial relationships among objects. Objects that are in the same site interact with each other. More than one site can reside in the same geographic location, but the objects contained in the different sites do not interact.

Examples

Using Sites to Organize Data

A drawing can have any number of sites. A designer may choose to use sites to organize parcel data. For example, in a multiphased subdivision, you could create a site for each phase of the subdivision and locate the parcels for each phase in a site specific to the phase. A designer may also choose to organize parcels in a site based on geographic location. This makes it easier to manage and work with a drawing that may contain several hundred parcels over a large area.

Topology

Topology is recognized in sites. When multiple objects exist in the same site, they have a spatial relationship. For example, when you change the location of a parcel segment that adjoins two parcels, each adjacent parcel automatically updates. This is shown in the following illustration.

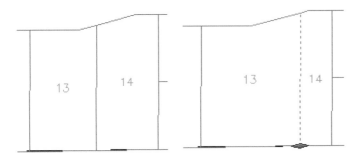

Furthermore, if you were to bisect a boundary parcel with an alignment from the same site, the parcel would automatically subdivide into two smaller parcels, one on each side of the alignment.

Object from Different Sites

Objects that belong to different sites are not affected by topology. For example, an alignment in one site may bisect a parcel in another site. Because the alignment and parcel are in different sites, the parcel does not automatically subdivide into two parcels.

Process: Creating Sites

The following steps show you how to create sites and then parcels from a boundary polyline object.

1   In Prospector, create the new site.

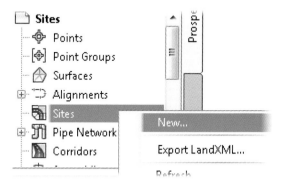

2   Enter site information.

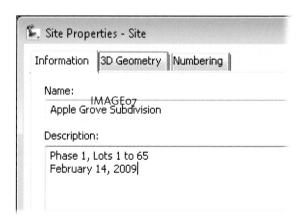

3   Create a parcel object from the boundary polyline. Enter parcel information.

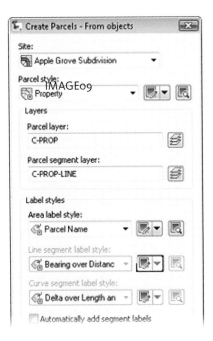

Guidelines for Creating Sites

Keep the following guidelines in mind when creating sites.

- When working with large amounts of data in a site, use representative naming conventions to help you organize your data.

- For good data management practices, use sites to organize parcel, alignment, feature line, and grading group data either geographically or by project phase.

- If you want design objects to interact with each other, assign them to the same site.

Exercise 01 | Create a Site

This exercise describes how to create a site for parcels.

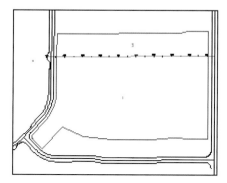

The completed exercise

1   Open \Site Design - Parcels\I_sites.dwg (Metric: M_sites.dwg).

2   In Toolspace, on the Prospector tab, right-click Sites. Click New.

3   In the Site Properties dialog box, for Name, enter **Apple Grove Subdivision**. Click OK.

4   In Prospector, click to expand Sites and Apple Grove Subdivision.

Note that a site can contain alignments, feature lines, grading groups, and parcels.

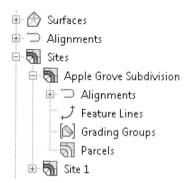

Next, you create a parcel object from the survey boundary polyline.

5   On the ribbon, Home tab, Create Design panel, click Parcel > Create Parcel from Objects.

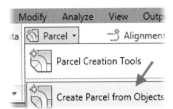

6   When prompted, select the blue property boundary in the drawing area. Press ENTER.

7   In the Create Parcels - From Objects dialog box:

- For Site, ensure that Apple Grove Subdivision is selected.

- For Parcel Style, ensure that Property is selected.

- Click OK.

8   In Prospector, click to expand Sites, Apple Grove Subdivision, Parcels. Click Property: 1.

If you cannot see the preview in the item view area, click the preview icon.

9   Under Apple Grove Subdivision, right-click Parcels. Click Show Preview. Click Property: 1.

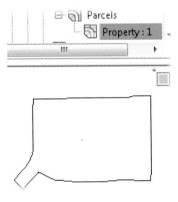

10   In Prospector:

- Click to expand Alignments, Centerline Alignments.

- Right-click Orchard Road. Click Move to Site.

11   In the Move to Site dialog box, for Destination Site, click Apple Grove Subdivision. Click OK.

The Orchard Road alignment is moved to the Apple Grove Subdivision site.

12   In the Apple Grove Subdivision site:

- Click to expand Alignments, Centerline Alignments.

- Notice the Orchard Road alignment.

- Notice that the parcel list in the Apple Grove Subdivision site needs refreshing.

13   Right-click Parcels. Click Refresh.

Notice that there are now two parcels in the site. This is an example of topology. After you move an alignment to the same site as the parcel it runs through, the alignment automatically subdivides the original parcel into two smaller parcels.

14   If you cannot see the two parcels in the drawing area, regenerate the drawing.

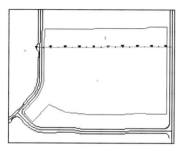

15   Close the drawing. Do not save the changes.

Lesson 17 | Creating Right-of-Way Parcels

This lesson describes how you create right-of-way (ROW) parcels.

You create a right-of-way parcel as one of your first tasks for designing a residential subdivision. After the right-of-way is created, you can subdivide the adjacent land parcels into individual lots. Parcels are dynamic objects that are defined by parcel segments. Furthermore, parcels in the same geographic location are usually created in the same site, and are therefore related topologically. This means that when you change the location of a parcel segment, any adjoining parcels automatically update.

The following illustration shows the parameters for a right-of-way parcel.

Create Right Of Way	
Parameter	**Value**
⊟ ☑ **Create Parcel Right of Way**	
Offset From Alignment	9.000'
⊟ ☑ **Cleanup at Parcel Boundaries**	
Fillet Radius at Parcel Boundary Intersections	9.000'
Cleanup Method	Chamfer
⊟ ☑ **Cleanup at Alignment Intersections**	
Fillet Radius at Alignment Intersections	9.000'
Cleanup Method	Chamfer

Objectives

After completing this lesson, you will be able to:

- Describe right-of-way parcels.

- List the guidelines for creating right-of-way parcels.

- Create a ROW parcel.

About Right-of-Way Parcels

You create a right-of-way parcel to identify a road allowance.

Definition of Right-of-Way Parcels

A right-of-way parcel represents the parcel of land for a road, also known as a road allowance.

Alignments and Parcels

When an alignment is in the same site as a parcel, and it passes completely through the parcel, the parcel automatically subdivides into two parcels, one on either side of the alignment. You use the Create ROW command to create a right-of-way parcel adjacent to the alignment. You specify offsets and can use fillet or chamfer cleanup options at the alignment intersections and the parcel boundaries.

Example

If you have a parcel that represents a boundary survey, and if that parcel is bisected with a road alignment object, the parcel splits into two. The Create ROW command then creates a right-of-way parcel for the road allowance based on the user inputting the right-of-way width (offset from the centerline).

Process: Creating ROW Parcels

The following steps show you how to create a ROW parcel.

1   Create a ROW parcel for the alignment road allowance separating Parcel 1 and 2.

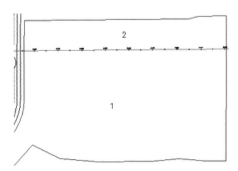

2   Modify parcel parameters such as the Offset and Cleanup methods.

Create Right Of Way	
Parameter	Value
☑ **Create Parcel Right of Way**	
Offset From Alignment	9.000'
☑ **Cleanup at Parcel Boundaries**	
Fillet Radius at Parcel Boundary Intersections	9.000'
Cleanup Method	Chamfer
☑ **Cleanup at Alignment Intersections**	
Fillet Radius at Alignment Intersections	9.000'
Cleanup Method	Chamfer

3   There are now four parcels, with the alignment road allowance parcels separating Parcel 1 from Parcel 2.

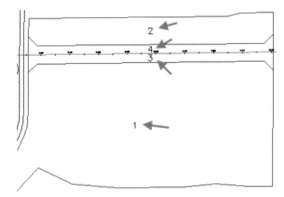

4   Use the Parcel Union command to join the two right-of-way parcels to form a single right-of-way parcel.

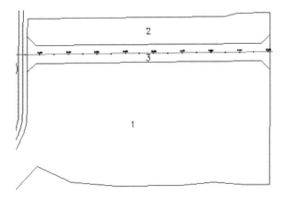

Guidelines for Creating ROW Parcels

Keep the following guidelines in mind when creating ROW parcels.

- The right-of-way functions like a narrow parcel, but it is not dynamically linked to the alignment. If you move or edit the alignment, you must delete the linework and rerun the command.

- When you run the Create ROW command, you are prompted to select one or more parcels. If an alignment is found on one of the edges of the selected parcels, a right-of-way is created in accordance with the supplied parameters.

Exercise 01 | Create ROW Parcel

In this exercise, you create a right-of-way parcel.

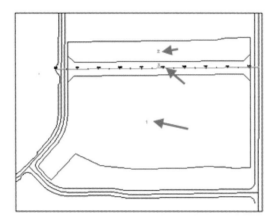

The completed exercise

1   Open *Site Design - Parcels\l_rightofway_parcels.dwg* (*Metric: M_rightofway_parcels.dwg*).

2   On the ribbon, Home tab, Create Design panel, click Parcel > Create Right of Way.

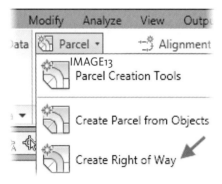

3 When prompted to select parcels, click to select both of the parcel labels. Press ENTER.

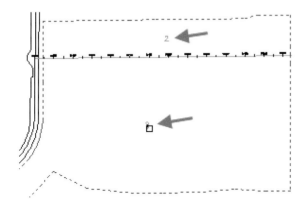

4 In the Create Right of Way dialog box, verify the following settings:

- Offset from Alignment is 30' (9 m).

- Fillet Radius at Parcel Boundary Intersection is 30' (9 m).

- Cleanup at Parcel Boundaries, Cleanup method is Chamfer.

- Click OK.

The two existing parcels were modified with a ROW buffer and chamfers at the parcel boundaries. Two additional parcels were created inside the ROW. Next, you combine these parcels, using the Parcel Union command.

5 On the ribbon, Home tab, Create Design panel, click Parcel > Parcel Creation Tools.

6   On the Parcel Layout Tools toolbar, select Parcel Union.

7   When prompted to:

* Select the destination parcel, click the label for Parcel 3.

* Select parcels, click the label for Parcel 4.

* Press ENTER.

8   Close the Parcel Layout Tools toolbar.

9   In Prospector, click to expand Sites, Apple Grove Subdivision, and Parcels. Note that there are three parcels on the list. If necessary, right-click Parcels. Click Refresh.

10   The final drawing contains the right-of-way parcel and a parcel to the north and south.

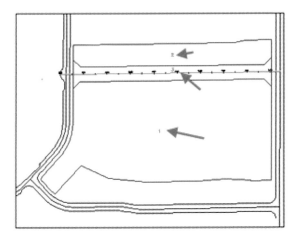

11   Close the drawing. Do not save the changes.

Lesson 18 | Creating Parcels

This lesson describes parcels and how to create parcels using the Parcel Layout tools or from objects. Often the first site is also the initial parcel from which all subsequent parcel design is generated. You can use parcels to create lots to precise specifications or to explore design alternatives.

The following illustration shows parcels created for a subdivision.

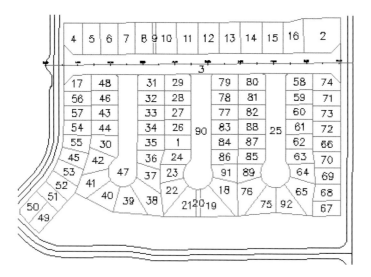

Objectives

After completing this lesson, you will be able to:

- Describe parcels and their properties.

- Describe the methods for creating parcels.

- List the guidelines for creating parcels.

- Create parcels using layout tools.

- Create parcels from objects that already exist in the drawing.

About Parcels

Parcels are created within a site. You use parcels to subdivide sites based on your project requirements, create lots to developer sizing specifications, or explore design alternatives.

Definition of Parcels

Parcel objects typically represent real estate parcels, such as lots in a subdivision or boundary surveys. You can also use a parcel to represent any feature with closed boundaries, such as bodies of water or soil regions. Parcels are defined by parcel segments, which can be lines or curves.

Additional Information

When working with parcels, keep the following information in mind:

- Parcel objects in the same site are related topologically and are, therefore, dynamic. This means that a change to one parcel creates related changes to the other adjacent parcels in the site.

- When you create a parcel, parcel area labels are automatically created. Parcel labels typically include a parcel number and area, but can include other information such as perimeter and address.

- You can create parcels using the Layout tools (Parcel Creation Tools) or from AutoCAD® objects.

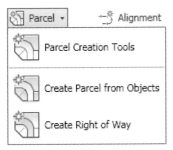

Examples

The following illustration shows parcels created for a subdivision.

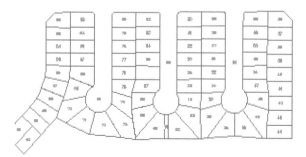

The dynamic nature of parcels in the same site enables you to quickly and efficiently make changes. For example, when you change the location of a parcel segment, any adjoining parcels automatically update. Or, if you create other parcel segments in the same site that bisect a parcel, the parcel subdivides into two other parcels. The new parcels contain labels appropriate to their location within the site and in the same style as the original parcel.

Methods for Creating Parcels

The first site is also the initial parcel from which all subsequent parcel design is generated. You can create parcels using the Parcel Layout Tools toolbar, or by using an object in your drawing. If you use the Parcel Layout Tools toolbar, you can select default style settings, create parcels by subdividing, resize parcels, and perform basic object editing.

Using Layout Tools

You can use commands on the Parcel Layout Tools toolbar to automatically subdivide larger parcels into smaller parcels. The two primary criteria used in laying out parcels are minimum road frontage and parcel area. If you want to control the parcel sizing using the area, you need to set the frontage criteria to a very small value. If you want to control the parcel sizing using the frontage distance, you need to set the area criteria to a very small value. If you use minimum area to size parcels, a larger parcel is automatically subdivided into several smaller parcels so that they meet the default area parameter. If you use a minimum frontage length to size parcels, the parcel area is ignored and parcels are created to satisfy the minimum frontage length criteria.

When creating parcels you can also specify frontage offset values and calculate parcels based on using the minimum frontage at the offset location.

For more information, see "Parcel Layout Toolbar" in Help.

The following illustration shows the Parcel Layout Tools toolbar with the parcel sizing parameters for default area and minimum frontage. In this example, because the minimum frontage is small, parcels are sized based on the area.

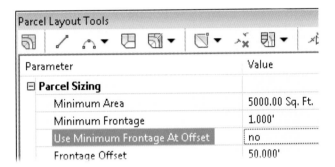

Process: Subdividing a Parcel

The following steps show you how to subdivide a parcel.

1   Open the Parcel Creation Tools toolbar.

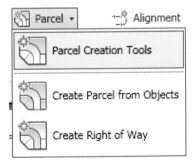

2   On the Parcel Layout Tools toolbar, click Slide Line - Create.

3   In the Create Parcels dialog box, select the site, parcel styles, and label styles.

4   Specify the frontage line and parcel segment direction.

5   Configure the parcel parameters.

Parameter	Value
Parcel Sizing	
Minimum Area	5000.00 Sq. Ft.
Minimum Frontage	1.000'
Use Minimum Frontage At Offset	no
Frontage Offset	50.000'

6   Preview the parcels, accept the results, and create other parcels.

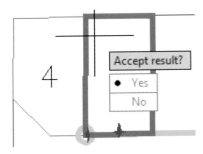

Using Objects

You can use AutoCAD objects such as lines, arcs, and polylines that form closed areas to create parcel objects. For example, a third-party organization such as a developer may provide the parcel fabric to the engineering company as AutoCAD lines and arcs. You could then convert the lines and arcs to parcel objects and thus be able to easily automate parcel labeling and create parcel closure reports.

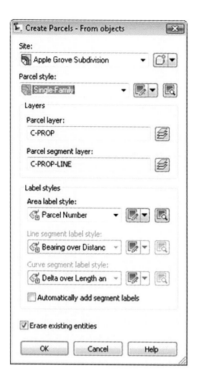

You launch the Create Parcels from Objects command on the ribbon, Home tab, Create Design panel.

Guidelines for Creating Parcels

Keep the following guidelines in mind when you create parcels.

- AutoCAD lines, arcs, and polylines should be well formed and represent closed areas prior to converting them to parcels.

- In order to use commands on the Parcel Layout Tools toolbar to subdivide a parcel, you first need to create a larger parcel to subdivide.

Example

You can create a subdivision based on an established property boundary. You create a parcel representing the property boundary either by selecting a drawing object that represents the property boundaries, or by using the Parcel Layout tools to create the individual boundary segments.

After you establish the initial parcel and other site elements, such as an alignment or right-of-way, new parcel segments, and therefore parcels, are automatically created when you use commands on the Parcel Layout Tools toolbar. The result is shown in the following illustration.

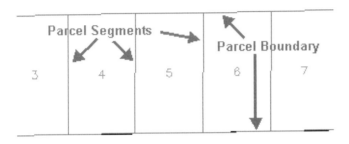

Exercise 01 | Create Parcels Using Layout Tools

In this exercise, you create parcels using layout tools. The two primary criteria used in laying out parcels are minimum road frontage and parcel area. If you want to control the parcel sizing using the area, you need to set the frontage criteria very small. If you want to control the parcel sizing using the frontage distance, you need to set the area criteria very small.

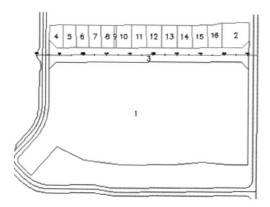

The completed exercise

1   Open \Site Design - Parcels\I_create_parcels_layout.dwg (M_create_parcels_layout.dwg).

2   Zoom in on Parcel 2, in the northern portion of the site.

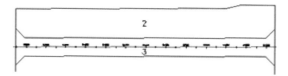

You use the layout tools to subdivide this parcel. First, you change the design settings for this command.

3   In Toolspace, click the Settings tab and do the following:

   • Click to expand Parcel, Commands.

   • Double-click CreateParcelByLayout.

4   In the Edit Command Settings - CreateParcelByLayout dialog box, expand Parcel Sizing and do the following:

- Click each of the parameters and review the settings.

- For Minimum Area, enter **5000 sq. ft. (470 sq. m)**.

- For Minimum Frontage, enter **1' (1 m)**.

- For Minimum Width, enter **1' (1 m)**.

- For Minimum Depth, enter **1' (1 m)**.

- For Use Maximum Depth, click No.

- For Multiple Solution Preference, click User Smallest Area.

- Click OK.

By setting the minimum frontage to a small value, the parcel sizing will be controlled by the parcel area criteria.

Next, you create the parcels.

5   On the ribbon, Home tab, Create Design panel, click Parcel > Parcel Creation Tools.

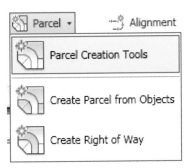

6   On the Parcel Layout Tools toolbar:

- Click the chevron to expand the toolbar.

- Click to expand Parcel Sizing. You should see the command settings you set previously in the command settings.

- Click to expand Automatic Layout.

- Verify that Automatic Mode is set to Off.

With the design layout criteria set, you now begin manually laying out parcels.

7   On the Parcel Layout Tools toolbar, click Slide Line – Create.

8   In the Create Parcels - Layout dialog box:

- For Site, click Apple Grove Subdivision.

- For Parcel style, click Single-Family.

- For Area label style, click Parcel Number.

- Click OK.

9   When prompted to select the parcel to be subdivided, in the drawing area, click 2, the parcel number.

10   If object snap is turned off, turn it on using OSNAP on the status bar.

11   At the Start Point on Frontage prompt, click the left side of the bottom line of the parcel using the Endpoint OSNAP. This is the start point for the frontage line.

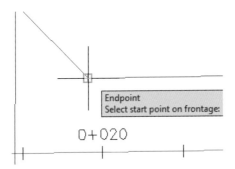

0+020

12 At the End Point on Frontage prompt, move the cursor to the right side of the bottom line of Parcel 2. Select the endpoint. Ensure that the frontage line is drawn along the south boundary for Parcel 2. This is the endpoint of the frontage line.

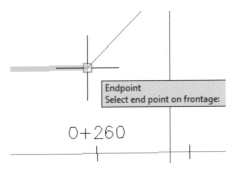

0+260

13 You are prompted to specify the angle used by the layout command for the right side of the new parcel. You can either enter the desired angle of **90** degrees, or you can click a start point and an endpoint to establish the angle.

At the Specify Angle at Frontage prompt, click the end of the chamfer line on the left edge of Parcel 2.

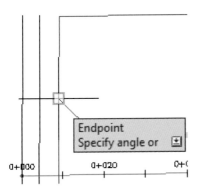

0+000   0+020   0+0

14 At the Second Angle Point prompt, click the upper corner of the left edge of the parcel.

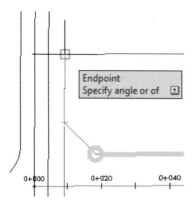

15 A parcel preview is displayed in the drawing area. You are prompted to accept the results.

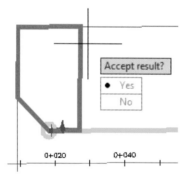

- Click Yes or press ENTER. The parcel is created.

16 You are prompted to accept the results to create the next parcel.

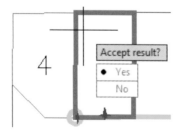

- Press ENTER to create the second parcel.

Chapter 05 | Site Design Parcels

17 Press ENTER three times to create three more parcels. Your drawing should look similar to the one that follows.

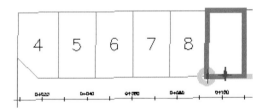

Next, you create a utility row with a width of 10' (3 m).

18 After you create Parcel 8, on the expanded Parcel Layout Tools toolbar:

- For minimum area, enter **1' (1 m)**.

- For minimum frontage, enter **10' (3 m)**.

The Slide Line - Create command is still active.

19 A parcel preview is displayed. Press ENTER to accept the results.

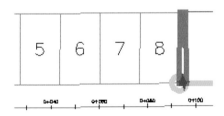

Next, you create the remaining parcels, this time using a different area.

20 On the Parcel Layout Tools toolbar:

- For Minimum Area, enter **6000 sq. ft. (500 sq. m.)**.

- For Minimum Frontage, enter **1 ft. (1 m)**.

- Under Automatic Layout, for Automatic Mode, click On.

A preview of the parcels is displayed.

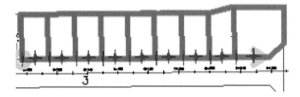

21   Press ENTER, or click Yes, to accept the results and create the parcels. Your parcel numbers may be different than shown.

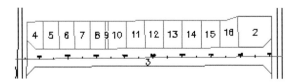

22   Close the Parcel Layout Tools toolbar.

23   In Prospector:

• Click to expand Sites, Apple Grove Subdivision, and Parcels.

• Right-click Parcels. Click Refresh.

• View the list of parcels you created.

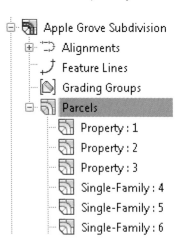

24 Right-click Property: 2. Click Properties.

25 In the Parcel Properties – Property: 2 dialog box, Information tab, for Object Style, click Single Family. Click OK.

26 The final drawing appears as follows. Your parcel numbers may be different than shown.

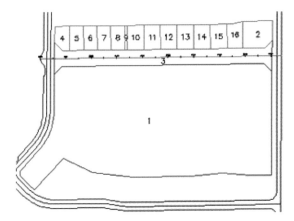

Close the drawing. Do not save the changes.

Exercise 02 | Create Parcels from Objects

In this exercise, you create parcels from objects that already exist in the drawing file. Designers often inherit a drawing that has existing objects that represent parcels. These objects can be lines, arcs, or polylines. For proper parcel analysis, labeling, and reporting functionality, you need to convert these to intelligent parcel objects.

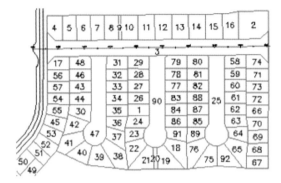

The completed exercise

1   Open \Site Design - Parcels\I_create_parcels_objects.dwg (M_create_parcels_objects.dwg).

2   In Toolspace, click the Prospector tab, and do the following:

- Click to expand Sites, Apple Grove Subdivision, Parcels.

- Right-click Property: 1. Click Zoom To. The drawing area zooms to Property: 1.

3   On the ribbon, Home tab, Layers panel, click Layer Properties.

4   In the Layer Properties Manager, thaw C-PROP- POLYLINES.

A set of closed polylines representing parcels is revealed.

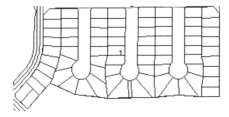

5   In the drawing area, click any one of the polylines that have been turned on. Right-click, click Select Similar.

     All of the polylines are selected.

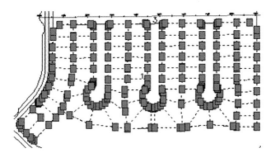

6   On the ribbon, Home tab, Create Design panel, click Parcel › Create from Objects.

     The polylines are already selected and the Create Parcels – From objects dialog box is displayed. This dialog box has several settings that should be inspected or modified. Any project can have multiple sites that contain alignments, feature lines, grading groups, and parcels. The process of creating parcels automatically creates a site. If a site already exists, you can assign the new parcels to the existing site.

7   In the Create Parcels from Objects dialog box:

     • For Site, click Apple Grove Subdivision.

     • For Parcel Style, click Single-Family.

     • For Area Label Style, click Parcel Number.

     • Clear the Automatically Add Segment Labels check box.

     • Select the Erase Existing Entities check box.

     Alternatively, you could change the command settings for CreateParcelFromObjects so that the parcel style is set to Single-Family automatically.

8   Click OK.

    The parcels are created.

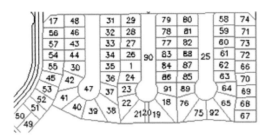

    Note: For this method of creating parcels, you must ensure that the polylines are connected at each vertex and carefully snapped. Any slivers between parcels result in phantom parcels.

9   In Prospector:

    - Click to expand Sites, Apple Grove Subdivision, Parcels.

    - If necessary, right-click Parcels. Click Refresh.

    - Click Apple Grove Subdivision.

    You see the overall site previewed in the item view area.

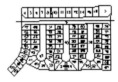

10 Under Parcels, click Property: 1.

    You see the preview for Parcel 1.

11   Right-click Single-Family: 73. Click Zoom To.

Parcel 73 is displayed in the drawing area. Your Parcel 73 may differ depending on the method used to pick the parcels during creation.

12   In Prospector, Sites, Apple Grove Subdivision, click Parcels. The item view window shows information about each parcel. Scroll right to view the area and perimeter of the parcels.

Name	Area	Perimeter
Property : 1	3657.52 Sq. Ft.	254.311'
Property : 3	52761.12 Sq. Ft.	1988.397'
Single-Family : 2	11365.69 Sq. Ft.	417.572'
Single-Family : 4	5000.00 Sq. Ft.	285.138'
Single-Family : 5	5000.00 Sq. Ft.	293.314'

You have created a total of 93 parcels, including 3 property parcels, and 90 single-family parcels.

13   The final drawing appears as follows.

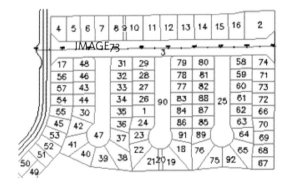

Close the drawing. Do not save the changes.

Lesson 19 | Editing Parcels

This lesson describes how to make changes to parcel objects using the editing and renumbering commands.

When parcels are created using commands from the Parcel Layout Tools toolbar, the segments for some parcels may need modification. Also, depending on the order in which parcels were created, you may need to renumber the parcels in the subdivision.

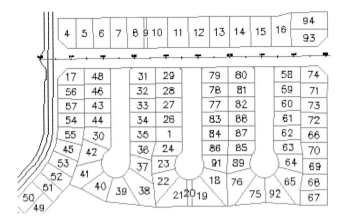

Subdivision that requires editing

Objectives

After completing this lesson, you will be able to:

- Describe the methods for editing parcels.

- Use the guidelines to help modify parcels.

- Edit parcels.

- Renumber parcels.

Methods for Editing Parcels

After you create parcels, you may need to edit them. Parcel geometry can be edited either graphically or by using commands on the Parcel Layout Tools toolbar. To modify parcel segment geometry graphically, you use the grips on ends of the parcel segments.

You can also renumber parcels within a site. Parcel numbering is independent of the site in which the parcels are created. This is useful when working on a large, multiphased subdivisions design project. Additionally, you may have instances where parcel numbers need to be the same.

Editing Parcel Geometry

There are two ways you can edit parcel geometry: graphically, or by using the commands on the Parcel Layout Tools toolbar. Graphical edits are used when you do not need to consider parcel area and frontage criteria for parcel editing. Use commands on the Parcel Layout Tools toolbar to edit parcels using area and frontage criteria.

Using Grips

For parcels created from AutoCAD lines, arcs, and polylines, you can grip edit a parcel object just as you would grip edit a polyline. This is shown in the following illustration.

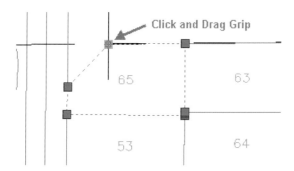

For parcels created using the layout tools, you can also use grips to modify parcel segments. However, the grips are slightly different from the previous example. Parcels segments created using commands from the Parcel Layout Tools toolbar have a grip on the parcel frontage. You can move the parcel segment, but not off the parcel frontage line. This is shown in the following illustration.

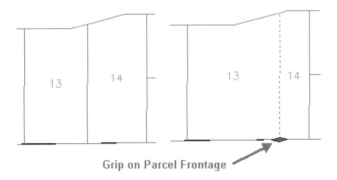

Grip on Parcel Frontage

Using Layout Tools

You can also edit parcels using the commands on the Layout Tools toolbar. Using the commands on the Layout Tools toolbar enables you to interact with the parcels at a higher level. For example, you can modify parcel segments to achieve a desired parcel area for an adjoining parcel.

Parcel editing commands on the Layout Tools toolbar are shown in the following illustration.

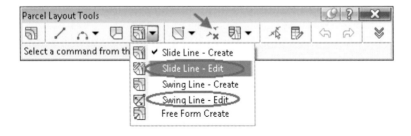

Renumbering Parcels

When parcels are created, they are assigned parcel numbers based on the order in which they were created. This can result in a parcel numbering sequence that is not suitable. You may also need to rename parcels. You can use the Renumber\Rename dialog box, as shown, to accomplish these tasks.

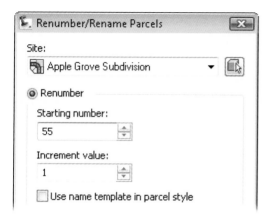

Guidelines for Editing Parcels

Keep the following guidelines in mind when you edit parcels.

- When graphically editing parcels, be aware of how the parcels were created. The grips you use for parcels created from objects are different from those for parcels created with commands from the Parcel Layout Tools toolbar.

- Use commands on the Parcel Layout Tools toolbar to edit parcels using design criteria such as parcel area and frontage.

Exercise 01 | Edit Parcels

In this exercise, you edit parcel segments.

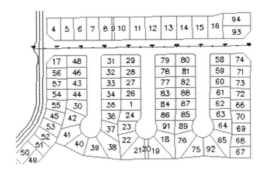

The completed exercise

1   Open \Site Design - Parcels\I_edit_parcels.dwg (M_edit_parcels.dwg).

First, you manually subdivide the larger parcel in the northeast section of the site.

2   Zoom in to Parcel 2, at the northeast section of the site.

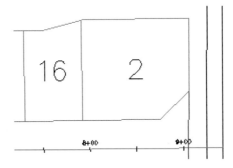

3   Press F3 to turn off object snaps.

4   On the ribbon, Home tab, Create Design panel, click Parcel > Parcel Creation Tools.

5   On the Parcel Layout Tools toolbar, click Add Fixed Line - Two Points.

6   In the Create Parcels - Layout dialog box:

- For Site, click Apple Grove Subdivision.

- For Parcel Style, click Single-Family.

- For Area Label Style, click Parcel Number.

- Click OK.

7   When prompted to specify a start point:

- Enter **MID**. Press ENTER to start the midpoint object snap.

- Move the cursor to a point approximately halfway up the left side of Parcel 2, and then click.

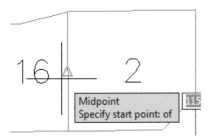

- When prompted to specify the next point, enter **PER** (perpendicular). Press ENTER.

- Move the cursor to the right parcel segment for Parcel 2, and then click.

8   Verify that your parcel is subdivided as shown.

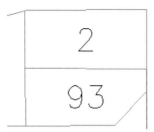

Next, you remove the unwanted polylines and parcel segments surrounding the road right-of- way.

9   On the Parcel Layout Tools toolbar, click Delete Sub-Entity.

10   When prompted to select a subentity to remove, click the parcel segments at the west end of the road right-of-way.

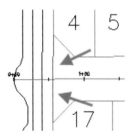

The parcel segments are deleted.

11   When prompted to select a subentity to remove, click the parcel segments at the west end of the road right-of-way.

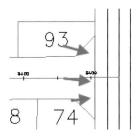

12 In the drawing area, click the parcel segments at the north end of the three cul-de-sac right-of-ways.

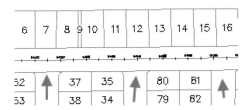

Notice the parcel segment in the northeast corner is now missing. Use the Add Fixed Line – Two Points on the Parcel Layout Tools toolbar to recreate this segment.

13 Close the Parcel Layout Tools toolbar.

The completed drawing appears as follows.

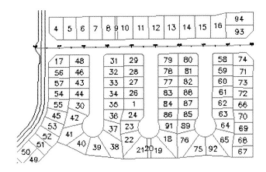

14 Close the drawing. Do not save the changes.

Exercise 02 | Renumber Parcels

In this exercise, you renumber parcels. Frequently, when you create parcels, the parcel numbers need to be rearranged so that they are numbered sequentially. To change the parcel numbering, you use the Renumber Parcels command.

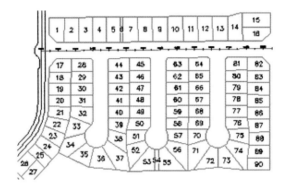

The completed exercise

1   Open *Site Design - Parcels\\l_renumber_parcels.dwg (M_renumber_parcels.dwg)*.

    Reviewing the current numbering scheme, you see that the parcel numbers are out of sequence. Next, you renumber the parcels.

2   On the ribbon, Modify tab, click Parcel.

    The Parcel tab is displayed.

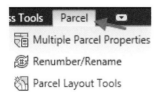

3   On the Modify panel, click Renumber/Rename.

4   In the Renumber/Rename Parcels dialog box, verify the following settings:

    • Site is Apple Grove Subdivision.

- Starting number is 1.

- Increment is 1.

- Click OK.

Next, you draw a line that crosses into each parcel, in the order desired for the parcels.

5   Click inside Parcel 4. Stretch the line northeast into Parcel 94. Click and then stretch the line to the east and then to the south as shown. Click.

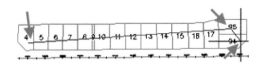

6   Press ENTER twice to end the command.

The parcels are renumbered.

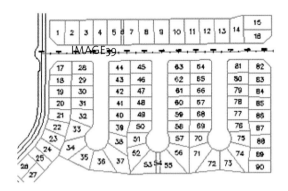

7   Repeat steps 3 through 6 to continue renumbering the southern section of the site as desired. The completed drawing appears as follows.

8   Close the drawing. Do not save the changes.

Lesson 20 | Labeling Parcel Segments and Creating Tables

This lesson describes how to label parcel segments with geometry data or tags, and how to create tables that show parcel data.

Parcel segment labels, tags, and tables provide important information to the reader. You can add parcel segment labels when you create the parcel, or you can add them later. When you add parcel segment labels, you can add labels that show the segment geometry or you can add tag labels. If you add tag labels to the parcel segments, you would then show the parcel geometry data in an associated table. When parcel segment geometry changes, associated labels and tables automatically update.

The following illustration shows a parcel with all segments labeled.

Objectives

After completing this lesson, you will be able to:

- Describe parcel segment labels.

- Describe parcel tags and parcel tables.

- Label a parcel and create a table.

About Parcel Segment Labels

Parcel segment labels can provide useful information about a parcel. When you create a parcel, you can add labels to parcel segments automatically. You can also add labels to existing parcels. When you add a parcel segment label, you can select to add a label to a single parcel segment or to all segments in the parcel. You can also set the style used for the segment lines and curves.

Definition of Parcel Segment Labels

Parcel segment labels provide useful information about the parts of the parcel. This information can include:

- Segment length.

- Bearing.

- Parcel segment number.

- Starting and ending coordinates.

Parcel Segment Label Options

After you create parcel segment labels, you can perform the following tasks:

- Change the label style.

- Reverse the direction of bearing information.

- Change the position of label information.

- Reset the label to its original properties.

- Delete the label.

Examples of Changing Label Appearance

The following illustrations show two examples of the types of changes you can make to the appearance of parcel segment labels.

The labels in the following illustration show south-to-west bearings and north-to-east bearings. You can use the Reverse Label command by right-clicking and accessing the shortcut menu if you want to change south-to-west bearings to north-to-east bearings.

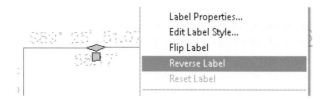

You can also use the Flip Label command from the same menu to flip the positioning of the bearing and distance components so that all of the bearings are on the outside and all of the distances are on the inside. The resulting labels are shown in the following illustration.

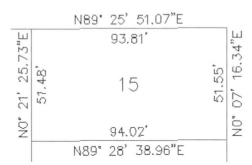

About Parcel Tags and Parcel Tables

Parcel tags and parcel tables enable you to organize data for a large number of parcels and parcel segments. They also help reduce the amount of clutter in your drawings.

Definition of Parcel Tags

Parcel tags are an alternative to parcel labels. Tags can be used for parcels and parcel segments. The parcel tag is cross-referenced in the accompanying parcel table. The tag value uniquely identifies the parcel or segment in the first cell of table rows.

Definition of Parcel Tables

A parcel table is a list of consolidated parcel information used for annotating a drawing. Parcel tables show the data associated with the tag. You create a table and place the table in a drawing for annotation. A table can contain line, curve, segment, or area information. The table information is dynamic, and therefore, updates made to parcels are reflected in the corresponding tables.

Key Points

Keep the following points in mind when working with parcel labels and tags:

- You can renumber parcel segment tag labels after they are created if you want to organize them sequentially.

- You can select individual labels in a drawing to include the label information in a table. You can also select label styles to use as tags and create a table containing information for all parcels and segments that use that style.

- You can show parcel data in a table as the design nears completion to make the diagram more readable.

- When you create tables, table tags that are part of the segment labels are made visible, and other information, such as the bearing, is hidden. You can use table tags to relate table information to objects in the drawing.

- When you create a line, curve, or segment table, the labels are removed from the parcel segments and replaced with a tag.

Examples

Parcel segments with tag labels are shown in the following illustration.

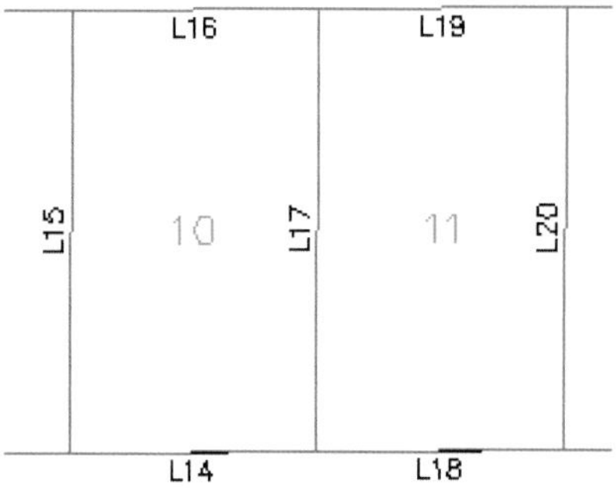

The corresponding table showing the data for the tags is shown in the following illustration.

Line #/Curve #	Length	Bearing/Delta	Ra
L24	42.51	S0° 21′ 25.73″W	
L20	42.66	S0° 21′ 25.73″W	
L13	19.23	S0° 21′ 25.73″W	
L11	53.91	S89° 25′ 51.07″W	
L8	54.66	S89° 25′ 51.07″W	
L5	42.77	N45° 06′ 21.02″W	
L4	28.43	S89° 25′ 51.07″W	
L36	50.97	S89° 25′ 51.07″W	

Parcel Line and Curve Table

Exercise 01 | Label Parcel Segments

In this exercise, you use a variety of labeling techniques for parcels. Creating a plat for the site requires that certain information about each parcel be displayed. These requirements vary depending on the jurisdiction, but generally the segment direction, lengths, and area of the parcels are required.

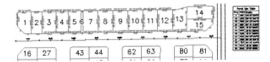

The completed exercise

1   Open *Site Design - Parcels\l_label_parcels.dwg (M_label_parcels.dwg)*.

2   In Prospector:

  • Click to expand Sites, Apple Grove Subdivision, Parcels.

  • Right-click Single-Family: 1. Click Zoom To.

   The drawing zooms in to Parcel 1.

3   Click the area label for Parcel 1.

   The parcels ribbon is displayed.

4   On the ribbon, Parcel: Single-Family: 1 tab, Labels & Tables panel, click Add Labels › Multiple Segment.

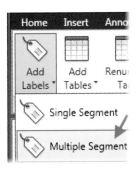

5   At the Select Parcel to be Labeled by Clicking on Area Label prompt, do the following:

  • Click the area label for Parcel 1.

  • At the Label Direction prompt, press ENTER to accept clockwise.

The parcel segments are labeled with the default parcel segment label style.

6   Press ENTER to end the command.

Next, you add labels to individual parcel segments.

7   On the ribbon, click the Annotate tab > Labels & Tables panel > Add Labels > Parcel > Add Parcel Labels.

8   In the Add Labels dialog box, for Line Label Style:

  • Click to expand Parcel Line Label Style, Bearing over Distance.

  • Click Distance.

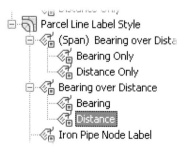

  • Click Add.

9   When prompted to select a point on entity, click the north, south, and east segments for Parcel 2.

Parcel 2 is labeled with distance labels.

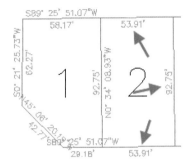

Next, you label the remaining parcels to the north of Orchard Road.

10 On the ribbon, Annotate tab, Labels & Tables panel, click Add Labels > Parcels > Multiple Segments.

11 Click the remaining parcel numbers, accepting Clockwise labels for each parcel, to the north of Orchard Road. Press ENTER when finished.

These parcel labels are dynamic and update when a segment is changed.

12 Note the labeled distances on the right-of- way for two adjacent parcels. Click the parcel segment between Parcels 2 and 3.

13 Click the grip. Move the mouse and click a new location for the parcel segment.

Notice the updated label distances for Parcels 2 and 3. You may need to regenerate the drawing to see the parcel labels update. Adding line labels can make the drawing difficult to read, especially in circumstances where the parcel segments are short.

Next, you add tag labels to the parcel segments.

14   Zoom in to Parcel 1 on the northwest corner of the site. Notice the overlapping segment labels in the southwest corner.

15   On the ribbon, Annotate tab, Labels & Tables panel, click Add Tables > Parcel > Add Line.

16   At the bottom of the Table Creation dialog box, click the No Tags Selected icon.

17   When prompted to select labels on lot lines:

   • Click the two overlapping labels you noted earlier.

   • Click the four parcel segment labels surrounding Parcel 6.

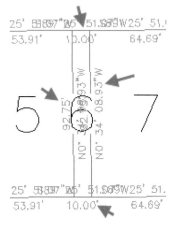

   • Click any other labels that appear crowded, near the west end of the parcels.

   • Press ENTER.

18   In the Create Table – Convert Child Styles dialog box, click Convert All Selected Labels Styles to Tag Mode.

19   In the Table Creation dialog box, click OK.

20 When prompted to select the upper-left corner, pick an empty location in the drawing to create the table.

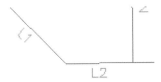

The parcel labels are converted to tags.

21 Note that a parcel segment table is inserted to the drawing.

Parcel Line Table		
Line #	Length	Direction
L1	42.77	N45° 09' 20.12"W
L2	28.18	S89° 26' 61.07"W
L3	10.00	S89° 26' 61.07"W
L9	42.17	S44° 45' 34.68"W
L4	92.75	N0° 34' 08.93"W
L5	92.75	N0° 34' 08.93"W
L11	61.60	N0° 34' 08.93"W
L6	10.00	S89° 26' 61.07"W
L7	18.50	S89° 26' 61.07"W
L8	48.11	S70° 50' 42.78"W
L10	20.49	N0° 07' 16.34"E
L12	52.05	N1' 45' 03.80'E

22 The final drawing appears as follows.

23 Close the drawing. Do not save the changes.

Chapter 06
Site Design Alignments

This chapter describes how you work with alignments as they relate to residential subdivision design projects. Alignments and road corridor models form the backbone for the subdivision design in that they usually set the grade for the parcels adjacent to the subdivision roads. The Creating Alignments from Objects lesson describes how you create alignments from polylines. The Labeling Alignments and Creating Tables lesson describes how you reduce the amount of annotation that appears on a subdivision plan by creating tag labels and showing the data in a corresponding table.

▶ **Objectives**

After completing this chapter, you will be able to:

- Create subdivision road alignment from polylines.

- Label horizontal alignments and create tables.

Lesson 21 | Creating Alignments from Objects

This lesson describes how you create subdivision road alignments from AutoCAD® entities such as lines, arcs, and polylines. Alignments are a critical component of all subdivision and roadway projects that have linear corridor design elements such as residential and collector roads. Alignments can also be used with creeks and rivers for floodplain analysis and channel design.

The following illustration shows two intersecting alignments:

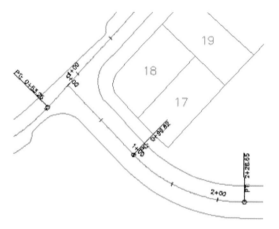

Objectives

After completing this lesson, you will be able to:

- Describe alignments and their properties.

- List the guidelines for creating alignments.

- Create alignments using objects.

About Alignments

The subdivision layout process is an iterative process where a developer, or the developer's engineer, strives to maximize the use of the land based on zoning, parcel layout, and road design criteria.

In many circumstances, the parcel outline for the subdivision is designed first, and then handed off to a designer who is tasked with designing the roads in the subdivision. Road designers often offset parcel right-of-way lines, polylines, and arcs to create the horizontal alignment geometry for the subdivision roads.

This geometry is converted to polylines, which are then used to create alignment objects for the subdivision roads.

Definition of Alignments

Alignments are a series of coordinates, lines, curves, and spirals used to represent the centerline of linear features such as roads, edges of pavement, sidewalks, and rights-of-way. Alignments can also be used to represent the centerline of a railway, channel, or stream.

Alignments Example

Horizontal alignments in subdivisions are usually not very complex and consist mostly of tangents and curves. In some instances lane tapers are modeled using alignments to create acceleration, deceleration, and turn lanes at intersection locations. Subdivision road centerline alignments are most often created by offsetting right-of-way lines by half the width of the right-of-way. Common commands such as trim, edit, extend, and fillet are used to create the alignment geometry from AutoCAD entities.

Once the geometry is in place, the Polyline Edit command can be used to join the lines and arcs together to form a continuous polyline representing the alignment. The direction of the polyline does not matter because once you create the alignment, you can reverse the direction of the alignment. You can also create alignments from AutoCAD line and arc entities.

Alignments for residential subdivision roads are shown in the following illustration.

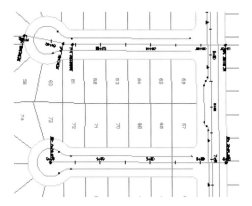

Guidelines for Creating Alignments

Keep the following guidelines in mind when you create alignments.

- The direction of the polyline is not important, as you can reverse the direction of an alignment during the alignment creation process. You can also reverse the alignment direction after it has been created.

- When you create an alignment from a polyline with no curves, or from lines, you can automatically add curves between the tangents.

- You can assign a value to the starting station of the alignment, which is the start point of the polyline, line, or arc. Alignment station reference points and base stationing values can be adjusted later.

- Alignments can either be independent or included in a site. Use alignments in a site if you want them to interact with other objects in the site, or if you want to use sites to organize the alignments.

Exercise 01 | Create Alignments from Objects

In this exercise, you create alignments using polylines.

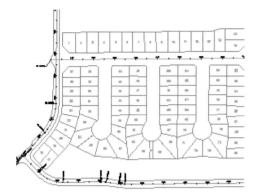

The completed exercise

You create an alignment from the polyline running south to north on the west boundary of the site, and from the polyline running west to east on the south boundary of the site.

1   Open *Site Design - Alignments\I_alignment_from_polyline.dwg*
    (*M_alignment_from_polyline.dwg*).

2   On the ribbon, Home tab, Create Design panel, click Alignment > Create Alignment
    from Objects.

3　At the Select Lines/Arcs or Polylines prompt, click near the left end of the polyline running west to east on the south end of the site. Press ENTER.

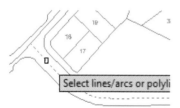

The stationing of the alignment begins at the end of the polyline closer to where you select it, or you can change the location.

4　Press ENTER to accept the alignment direction (west to east).

5　In the Create Alignment from Polyline dialog box:

- For Name, enter **Oak Street**.

- For Site, click <None>.

- For Type, ensure Centerline is selected.

- For Alignment Style, click Proposed.

- For Alignment Label Set, click All Labels.

- Under Conversion Options, clear the Add Curves Between Tangents check box.

- Under Conversion Options, click Erase Existing Entities. This option removes the original polyline once the alignment is created.

6　On the Design Criteria tab:

- Notice that you can assign a design speed and use criteria-based design. This means that you can check alignments for minimum curvature and assign superelevation from design criteria files.

- Click OK.

The alignment is created and labeled.

7   Repeat steps 4 and 5 for the road running south to north on the west side of the site. This
    road is called 8th Avenue, and the direction of stationing increases to the north.

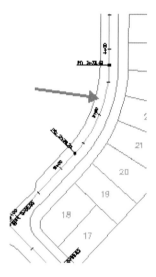

8   In Toolspace, on the Prospector tab:

    • Click to expand Alignments, Centerline Alignments, 8th Avenue.

    • Notice that each alignment can have multiple profiles, profile views, and sample
      line groups.

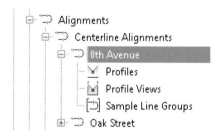

    • Right-click 8th Avenue. Click Properties.

9   In the Alignment Properties - 8th Avenue dialog box, Information tab, for Object Style, click Existing. Click OK.

    The alignment display updates.

10 Repeat steps 7 and 8 to create the south to north Oak Street alignment.

    Next, you view the object styles in the Settings tab.

11 In the Settings dialog box:

    • Click to expand Alignment, Alignment Styles.

    • Right-click Existing. Click Edit.

12 In the Alignment Style - Existing dialog box, on the Display tab, review the visible component settings.

    Note that only the Line, Curve, and Spiral components are visible and are set to the C- ROAD-CNTR layer, which is red.

    Also notice the Warning Symbol component. This is used to display a warning in the drawing area if you have used criteria-based design and violated the design criteria.

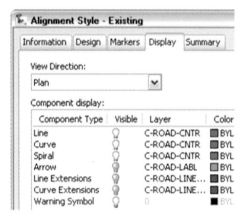

13 Click Cancel.

14 Repeat steps 11 and 12 and review the proposed alignment style. Note the alignment style differences.

    Next, you preview the alignment objects in Prospector.

15 In Prospector: Click to expand Alignments, Centerline Alignments.

- Click Oak Street. In the item view area you should see a preview of the alignment.

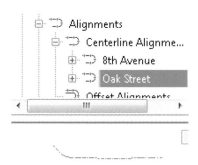

16 If you do not see the preview, ensure that the preview icon at the top of Prospector is on. Right-click Alignments. Ensure Show Preview is enabled.

17 The finished drawing appears as follows.

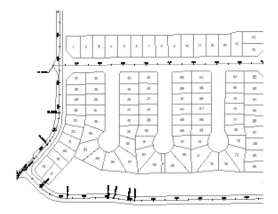

18 Close the drawing. Do not save the changes.

Lesson 22 | Labeling Alignments and Creating Tables

This lesson describes how you label horizontal alignments and create tables that show the alignment data. Horizontal alignments are made up of segments, which are lines, arcs, or spirals. There are a number of powerful labeling tools in Civil 3D® for labeling horizontal alignment geometry, either on the alignment itself or in a table.

When you edit or change an alignment, associated labels and tables automatically update to reflect the new alignment geometry.

The following illustration shows alignment geometry with segment labels.

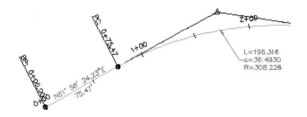

Objectives

After completing this lesson, you will be able to:

- Describe alignment tag labels.

- List the guidelines for creating alignments and tables.

- Label alignments and create a table.

About Alignment Tag Labels

When plans become difficult to read because of too many geometry alignment labels, you can create tag labels for the alignment segments and show the geometry in a corresponding table.

After adding tag labels, you create an alignment table that references the tags. You can create a line, curve, spiral, or segmental table that shows the geometry for the entire alignment. The table can be dynamic. When you edit the horizontal alignment or change the station reference point, the data in the table automatically updates to reflect the new geometry.

To create tag labels, you select a tag label style from the Add Labels dialog box as shown in the following illustration.

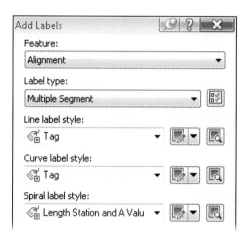

Definition of Alignment Tag Labels

There are a number of different label types that you can add to a horizontal alignment. To label alignment geometry, you can either label single segments or multiple segments. The multiple segment option enables you to label all segments for the entire alignment. When you choose the label type, you can then specify the corresponding label style.

An alignment with tag labels is shown in the following illustration.

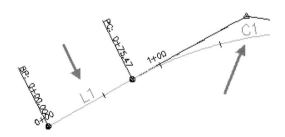

Tag and Table Example

The table in the following illustration displays segment numbers (tag label) and its associated details.

Huckleberry Hill				
Number	Radius	Length	Line/Chord Direction	A Value
L2		101.87	S51° 32' 01.01"E	
L3		156.94	S69° 41' 15.00"E	
L1		75.47	N61° 58' 24.23"E	
C1	308.23	196.32	N80° 13' 11.61"E	
C2	373.38	77.20	S75° 36' 38.00"E	

Guidelines for Labeling Alignments

Keep the following guidelines in mind when labeling alignments.

- Use the alignment command settings to specify default label styles for alignment labeling.

- Use tag labels and tables to simplify the appearance of a drawing. Note that when you create a table for an alignment that already has labels, the labels are automatically converted to tag labels.

Exercise 01 | Label Alignments and Create a Table

In this exercise, you create alignment tag labels and an alignment table.

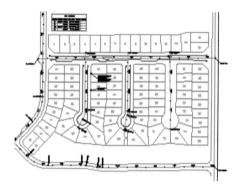

The completed exercise

1   Open *Site Design - Alignments\I_label_alignments.dwg (M_label_alignments.dwg)*.

First, you review the labels.

2   In Toolspace, Settings tab:

- Click to expand Alignment, Label Styles.

- Review the variety of label styles. You can create or modify styles in any of these categories. You can also individually label segments of alignments or points either on or offset from the alignment.

- Click to expand Label Sets.

- Right-click Label Sets. Click New.

3   In the Alignment Label Set - New Alignment Label Set dialog box:

- Click the Labels tab. Review the choices under Type. These are the types of alignment labels you can include in a label set.

- With Type set to Major Stations, click Add.

The Parallel with Tick Major Station label style is added to the label set.

- Change Type to Minor Stations. Click Add.

The Tick Minor Station label style is added to the label set. Note that you can specify the increment for the major and minor station labels. You can build your own customized label set.

- Click Cancel.

Alignment label sets are assigned to an alignment when you create it.

4   On the ribbon, Home tab, click Alignment > Alignment Creation Tools.

5   In the Create Alignment - Layout dialog box:

- Review the Alignment Label Set list. When you create an alignment by layout, you should specify the alignment label set.

- Click Cancel.

6   In the drawing area, select the 8th Avenue alignment. This is the north – south running alignment on the west side of the subdivision. Right-click and then click Edit Alignment Labels.

7   In the Alignment Labels dialog box:

- Notice that you can import a predefined label set.

- For Major Station Increment, enter **50' (50 m)**.

- For Minor Station Increment, enter **10' (20 m)**.

- Click in the Station Equations row, click Delete (red 'X').

- Click in the Profile Geometry Points row, click Delete (red 'X').

- Click Save Label Set.

- In the Alignment Label Set dialog box, Information tab, for Name, enter **Maj (50) Min (10) and GP**.

- Click OK twice.

The labeling of 8th Avenue has changed. You have also created a new label set.

Next, you add alignment station and offset labels.

8   In the drawing area, zoom in and click the 8th Avenue alignment.

9   On the contextual ribbon select Add Labels > Add Alignment Labels.

This is the same Add Label dialog box used for other features.

10 In the Add Labels dialog box:

- For Label Type, click Station Offset - Fixed Point.

- For Station Offset Label Style, click Station Offset and Coordinates.

- Click Add.

This command places a label that shows a station and offset from an alignment, and the coordinates of that point, for a location such as a fire hydrant or parcel boundary.

11 At the Select an Alignment prompt, select the Apple Ave centerline alignment.

Apple Ave is the western cul-de-sac alignment.

12 Move the mouse near the west endpoint of the parcel boundary between parcels 43 and 44.

13 When prompted to Select Point, use the endpoint object snap and select a parcel corner.

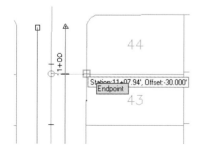

A station offset is created.

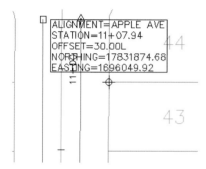

14 In the Add Labels dialog box, for Station Offset Label Style, click Station and Offset. Click Add.

15 At the Select Alignment prompt, click Apple Ave.

16 At the Select Point prompt, move the cursor south and near the west endpoint of another parcel segment. Snap to the end point.

17 Press ESC. A different station offset label is created.

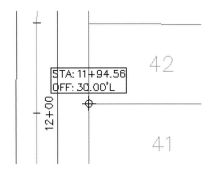

Next, you experiment with the label dragged state display properties.

18 In the drawing area, select a station and offset label.

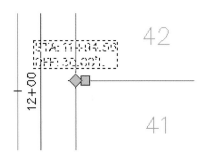

19 Hover the mouse over each grip. Notice the tooltips.

You use the Move Point Being Labeled grip (diamond shape) to reposition the label and the label point.

You use the Drag Label grip (square shape) to drag the label away from the point it is labeling. A pointer is added to the point label to show the point being labeled.

20 Experiment with the Slide Label and Drag Label grips on the Station and Offset labels you created.

Next, you label the alignment segments.

21 In the Add Labels dialog box:

- For Label Type, click Single Segment. This command is used to label an individual segment (line, curve, or spiral) of an alignment.

- For Line Label Style, click Bearing over Distance.

- Click Add.

22 At the Select Point on Entity prompt, click a line segment on the Orchard Road alignment. This is the alignment that intersects the three cul-de-sac alignments.

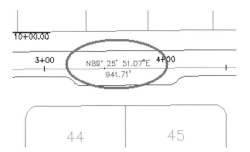

23 Press ESC.

24 Click the line label on the alignment and experiment with the grips.

Next, you create tag labels on the alignment.

25 In the Add Labels dialog box:

- For Label Type, click Multiple Segment.

- For Line Label Style, click Tag.

- For Curve Label Style, click Tag.

- Click Add.

26 Under Select Alignment, click 8th Avenue.

This is the south-to-north running alignment on the west side of the site.

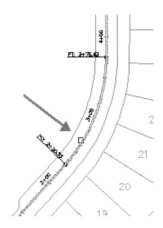

The tag labels are created.

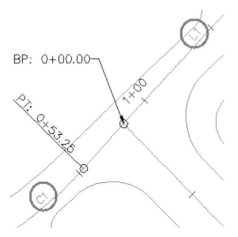

- Press ENTER.

- Click Close to close the Add Labels dialog box.

Next, you create a table to show the alignment data.

27 In the drawing area, select the 8th Avenue alignment.

28   On the contextual ribbon, select Add Tables > Add Segments.

29   In the Alignment Table Creation dialog box, for Select Alignment, click 8th Avenue. Click OK.

30   At the Select Upper Left Corner prompt, click in the drawing to create the table.

The table is created showing the alignment data for the 8th Avenue alignment.

8th Avenue				
Number	Radius	Length	Line/Chord Direction	A Value
C1	193.57	53.25	N51° 12' 50.83"E	
L1		176.99	N43° 20' 01.10"E	
C2	193.57	145.19	N21° 50' 43.41"E	
L2		450.42	N0° 21' 25.73"E	

31   The finished drawing appears as follows.

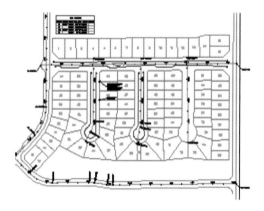

32   Close the drawing. Do not save the changes.

Chapter 07
Site Design Profiles

This chapter describes how you create and present existing and design profiles for subdivision roads. Existing ground profiles typically represent the current topography below the location of the proposed subdivision alignments. Design profiles, or layout profiles, are used to set the design and construction grades for the roads in a subdivision. You present profile data by creating profile views.

The Creating Surface Profiles and Profile Views lesson describes how you create surface profiles and profile views. The Creating Layout Profiles lesson describes how you create design layout profiles using commands on the Profile Layout Tools toolbar. The Editing Profile Geometry lesson describes how you edit layout profile geometry graphically, by editing the table in Panorama, or by using the entity creation commands. Finally, the Labeling Profiles and Profile Views describes how label profiles and profile views to ready the drawing for review.

▶ Objectives

After completing this chapter, you will be able to:

- Create surface profiles and profile views.

- Create profiles by layout and use transparent commands to create profile geometry.

- Edit profile geometry.

- Label profiles and profile views.

Lesson 23 | Creating Surface Profiles and Profile Views

This lesson describes how to create surface profiles and profile views. Surface profiles are created for alignments and typically show the nature of the existing terrain along the alignment. Profile views are the grid objects that show surface profile and other types of profile data.

Surface profiles are dynamic objects that automatically update if either the horizontal alignment geometry changes or the surface changes. This makes it very easy to adjust the horizontal alignment to best match the existing terrain.

A profile view with a surface profile is shown in the following illustration.

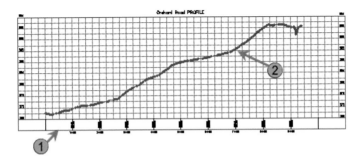

① Profile view

② Surface profile

Objectives

After completing this lesson, you will be able to:

- Describe surface profiles.

- Explain profile views.

- List the guidelines for creating profiles.

- Create a profile from a surface and create a profile view.

About Surface Profiles

After you create an alignment, the next step in the road design process is to generate the surface profile. The surface profile helps the designer determine the layout of the design profile. In many instances, engineers strive to design road profiles to match the surface profiles as closely as possible to minimize the amount of earthworks on a project. Surface profiles are dynamic profiles and automatically update when you edit the alignment geometry or station data, or change the surface.

Definition of Surface Profiles

Surface profiles are objects used to represent terrain data along a horizontal alignment. Surface profiles are displayed in profile views.

Data Points

Surface profile data points are generated wherever the alignment intersects a TIN line. The following illustration shows the locations where the data points are sampled when a surface profile is created for an edge of pavement alignment.

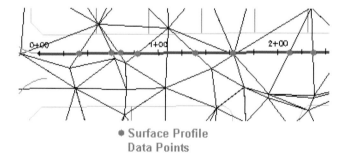

● Surface Profile
Data Points

Profile Editor

Profile data is displayed in the Profile Editor. The surface profile data station and elevation pairs are shown in the following illustration. These are the locations where the alignment intersects a surface triangulation line.

3	0+82.98'	272.123'	-0.26%
4	1+07.34'	272.251'	0.53%
5	1+30.00'	272.896'	2.85%
6	1+58.17'	274.311'	5.02%

Example

The following illustration shows a surface profile with an existing elevation change from west to east along the centerline of an alignment.

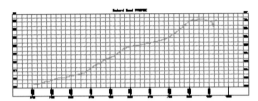

About Profile Views

You need a profile view to display surface profile and other types of profile data.

Definition of Profile Views

A profile view is a graph that includes a grid, X and Y axes, and data bands that displays profile view data. The X axis represents horizontal distance along the selected horizontal alignment or other linear feature. The Y axis represents elevations. You can configure data bands to contain annotations such as elevation data, stations, and cut/fill depths.

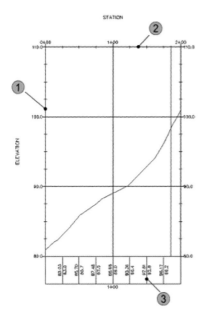

(1) Y axis

(2) X axis

(3) Data bands

Profile View Style

You control the appearance of a profile view using profile view styles. The profile view style also controls the following:

- Vertical exaggeration of the profile view object.

- Major and minor grid spacing.

- Axes labeling.

- Profile view title annotation.

- Colors, linetypes, and lineweights of profile view components.

You can customize the components and appearance of the graph. You can change the display elements of a profile view, such as title format, axis annotation, and grid appearance. You can set the view to clip grid lines above the profile line or to hide grid lines.

Examples

In the first illustration, notice that the surface profile is at a significant grade. The profile view style shows the profile data at a 10:1 vertical exaggeration. This may result in difficulties fitting the profile data in the profile portion of a plan and profile construction drawing.

For steep terrain, a 10:1 vertically exaggerated profile view would not show the entire profile within a paper space viewport. To address this issue, the designer would create and assign a profile view style that reduces the vertical exaggeration to 5:1. The profile view with the new profile style assigned is shown in the second illustration.

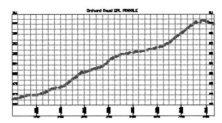

Profile view with 10:1 vertical exaggeration

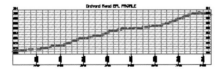

Profile view with 5:1 vertical exaggeration

Guidelines for Creating Profiles

Keep the following guidelines in mind when creating profiles.

- For surface profiles to be accurate, you must ensure that the surface is accurate. This means using breaklines and excluding points that do not represent the lay of the land from the surface.

- When you move the profile view, profile data in the profile view moves with the profile view.

- If you change the vertical exaggeration of a profile view by assigning a new profile view style, profile data in the profile view updates accordingly.

Exercise 01 | Create a Surface Profile and a Profile View

In this exercise, you create a surface profile using an alignment and an existing surface. You also create a profile view.

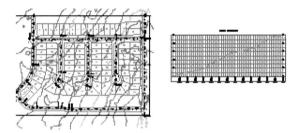

The completed exercise

1   Open *Site Design - Profiles\I_surface_profiles.dwg* (*M_surface_profiles.dwg*).

    First, you display the existing ground surface as contours.

2   In Toolspace, on the Prospector tab:

- Click to expand Surfaces.

- Right-click Existing Ground. Click Surface Properties.

3   In the Surface Properties - Existing Ground dialog box, Information tab, for Surface Style, click Contours 1' and 5' Background (Contours 1m and 5m Background). Click OK.

    The surface, with contours showing, is displayed. Note that this step is not necessary to create a surface profile.

4   In the drawing, select the Orchard Road centerline alignment. Orchard Road is the west- to-east road that intersects with the three cul- de-sac roads.

5   On the ribbon, Launch Pad panel, click Surface Profile.

6   In the Create Profile from Surface Dialog box:

-   For Alignment, click Orchard Road.

-   Under Station Range, do not make changes.

The entire alignment will be sampled.

-   Under Select Surfaces, Existing Ground is selected because it is the only surface in the drawing.

-   Click Add.

-   Under Profile List, stretch the Name column to see the profile name.

At this point, you can add more profiles by changing the alignment or adding offset alignments to Orchard Road by selecting the Sample Offsets check box. You can also create profiles without creating a profile view in the drawing. A profile and a profile view are separate objects, as shown in the Prospector tab.

Next, you create the profile view.

7   In the Create Profile from Surface dialog box, click Draw in Profile View.

The Create Profile View wizard opens. Note the various settings in the steps as you move through the wizard.

8   On the General page, for Select Alignment, select Orchard Road. Click Next.

You can change the range of the alignment to view.

9   On the Station Range page, click Next.

10   On the Profile View Height page, click Next.

11   On the Profile Display Options page, click Next.

The data bands refer to the information displayed above or below the profile view. Data bands can show many different types of data including profile, alignment, pipe network and superelevation.

12   Click Next to display the Profile Hatch Options.

You can hatch areas between multiple profiles to graphically show areas of cut and fill. In this exercise, because there is only one profile, hatching is not possible.

13   Click Create Profile View.

14   At the Select Profile View Origin prompt, click to the right of the plan view of the site to set the origin of the profile view.

The profile view is created in the drawing.

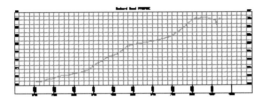

15   Zoom in to the profile view to review the surface profile data.

The profile view shows the existing elevation change from west to east along the centerline of Orchard Road. The data band shows the station, and the existing and finish elevations, which are the same because there is no design profile. You can show multiple profiles in the same view, which can be very helpful when comparing the existing and the design profile elevations and grades.

Next, you examine profiles and profile views in Prospector.

16   In Prospector:

- Click to expand Alignments, Centerline Alignments, Orchard Road.

- Click to expand Profiles and Profile Views.

Recall that the alignment and the profile are dynamically linked.

17   At the command line, enter **VPORTS**. Press ENTER.

18   In the Viewports dialog box, Named Viewports tab, for Named Viewports, click Plan & Profile. Click OK.

19   If necessary, adjust the zoom and pan so that the Orchard Road plan view is in the top viewport and the profile view is in the bottom viewport.

20   In the top viewport, select the Orchard Road alignment.

21 Click a grip and move the mouse slightly to the north. Click again to move the alignment.

Notice the profile change in the lower viewport.

22 At the command line, enter **UNDO** to undo the alignment change. Press ENTER.

23 Enter **VPORTS**. Press ENTER.

24 In the Viewports dialog box, on the New Viewports tab, under Standard Viewports, select Single. Click OK.

The drawing for the completed exercise appears as follows.

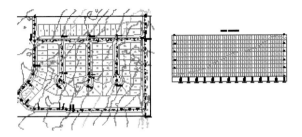

25 Close the drawing. Do not save the changes.

Lesson 24 | Creating Layout Profiles

This lesson describes how you create design layout profiles using commands on the Profile Layout Tools toolbar. It also describes how you use transparent commands to help you create profile geometry using grades, stations, and elevations.

After the designer creates a surface profile, the next step in the road design process is to create a layout profile. The layout profile represents the design profiles and consists of tangents and vertical curves. You create layout profiles by using commands on the Profile Layout Tools toolbar.

Objectives

After completing this lesson, you will be able to:

- Describe layout profiles.

- Explain when to use transparent commands.

- List guidelines for creating layout profiles.

- Create a layout profile.

Chapter 07 | Site Design Profiles

About Layout Profiles

Unlike dynamic surface profiles, layout profiles are static and do not update with changes to the alignment. You can create layout profile geometry by converting existing AutoCAD® lines and splines. You can also create layout profiles by inputting stations, elevations, and vertical curve data in a tabular editor. There are a number of useful commands available on the Profile Layout Tools toolbar that help you create a layout profile.

The following illustration shows a layout profile in the profile view. The surface profile is also present in the profile view.

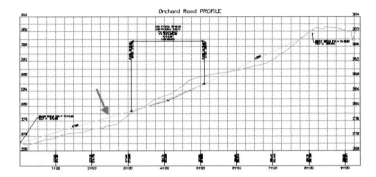

Definition of Layout Profiles

Layout profiles represent the finished vertical alignment that final road construction and grading is based on. The layout profile is also known as the vertical alignment or design profile and consists of tangents and vertical curves.

Profile Layout Considerations

You can lock a PVI at a specific station and elevation so that it cannot be moved by edits to adjacent entities. This is often done in intersections, where a side road profile is locked to the crown and edge of pavement elevations of the main road. If you unlock a dynamic PVI for one road in an intersection, you break the link to the profile of the other road.

After you create and display the surface profile in the profile view, you then create the layout profile, which represents the design of the vertical alignment for the road.

You define your layout by selecting points of vertical intersection (PVIs), which are connected using tangents either with or without curves. If you are drawing tangents with curves, you can choose standard curve settings to create an initial curve design. You can edit the curve values and other profile layout properties later to generate a more precise design.

When you create a layout profile, you select a profile style and label set. You can later change the profile style and label set to suit the phase of the road design. For example, you may want fewer labels as you lay out the profile and more labels when you are preparing the design for construction documents.

Profile Layout Examples

In many instances a designer needs to reproduce design profile data that already exists on a hard-copy plan. The hard-copy plan would show PVI stations, PVI elevations, vertical curve lengths, and tangent grades.

You can use the Insert PVIs Tabular command on the Profile Layout Tools toolbar to create a layout profile, as shown in the following illustration.

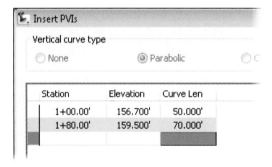

The following examples show a profile view and surface profile before and after the layout profile is created.

The following illustration shows a profile view with a surface profile. When you create the layout profile, you can reference stations, elevations, and grades in the profile view using the commands on the Transparent Commands toolbar.

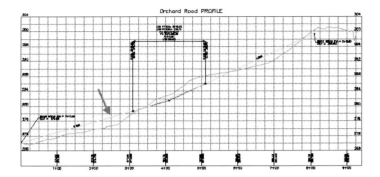

In this illustration, the layout profile is drawn in the profile view with PVIs and vertical curves. You can configure the profile to automatically place labels at standard points of reference as you create it.

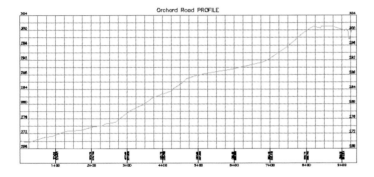

About Transparent Commands

Transparent commands are designed to simplify drawing tasks. Transparent commands can be used to reference locations in both plan and profile views. You use the profile transparent commands to create profile geometry by referencing stations, elevations, and grades.

The Transparent Commands toolbar is shown in the following illustration.

Definition of Transparent Commands

Transparent commands are commands that are issued from within another command. You can use transparent commands to locate objects based on known information. These locations are usually identified in the context of a larger operation, such as while drawing a line within a parcel. You can use transparent commands when creating any type of Civil 3D® object such as points, parcels, profiles, or alignments.

For more information, see "Civil Transparent Commands" in Help.

Example of Using a Transparent Command

You can use the Profile Station Elevation Transparent command to create or move a PVI to a known station and elevation in a profile view.

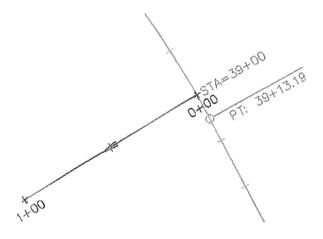

Guidelines for Creating Layout Profiles

Keep the following guidelines in mind when you create layout profiles.

- Layout profiles are static, not dynamic. Therefore, when you edit horizontal alignment geometry, you also need to edit the layout profile.

- Layout profiles must be created in the direction of increasing change.

Exercise 01 | Create a Layout Profile

In this exercise, you create a layout profile.

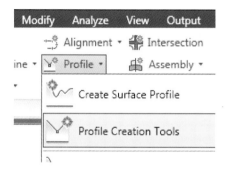

The completed exercise

1   Open *Site Design* - Profiles*l_layout_profiles.dwg* (*M_layout_profiles.dwg*).

2   On the ribbon, Home tab, Create Design panel, click Profile > Profile Creation Tools.

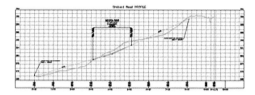

3   At the Select Profile View prompt, in the drawing area, click the Orchard Road profile view.

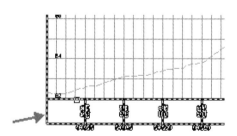

4   In the Create Profile - Draw New dialog box:

- For Name, enter **Orchard Road - Design Profile**.

- Notice the profile style is set to Design Profile and the profile label set is set to Complete Label Set.

- Click the Design Criteria tab. Notice that you can use criteria-based design to ensure the design profile meets the minimum requirements.

- Click OK.

The Profile Layout Tools toolbar appears. The following steps use these tools to create the design profile directly in the profile view in the drawing. You specify curve settings and place tangents to create the proposed vertical layout for Orchard Road. If you make a mistake as you design the profile, you can use the Undo or Redo buttons on the right side of the toolbar.

First, inspect the vertical curve settings.

5   On the Profile Layout Tools toolbar, click Curve Settings.

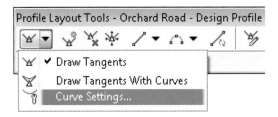

6   In the Vertical Curve Settings dialog box:

- Review the options for the vertical curve settings.

- Under Crest Curves, for Length, enter **100 (30 m)**.

- Under Sag Curves, for Length, enter **100 (30 m)**.

- Click OK.

Parabolic curves are most commonly used for vertical road curves, but you can modify this setting. Both crest and sag curves can either be designed using a specified length or a K value.

7 On the Profile Layout Tools toolbar, click Draw Tangents with Curves.

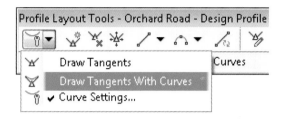

There are multiple methods for creating layout profiles. The Draw Tangents with Curves method is very easy to use. You simply click in locations where you would like a PVI (point of vertical intersection).

8 At the Specify Start Point prompt, use the Endpoint object snap and click the western end of the surface profile.

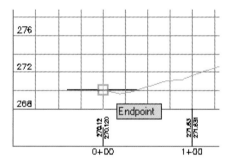

As you move the mouse away from the start point, the profile stretches, awaiting the position of the next PVI. Each point that you click will be a PVI, the end of one tangent and beginning of the next. Civil 3D automatically fits the curve to transition between the two tangents.

Normally, the goal is to position these points so that the areas of cut and fill nearly balance, although the site grading near the alignment may influence the final balance. The profile being designed is the first attempt at the finished vertical grade. Be careful when using object snaps during this procedure.

9 Zoom out to give a better view of the profile.

10   Using your best judgment, click approximately four or five evenly spaced locations for PVIs along the profile view, ending with a click at the endpoint of the profile.

Make sure you do not select PVI locations too close together. Remember that the specified vertical curve length is 100' (30 m). This means that you need a minimum distance of 100' (30 m) between consecutive PVIs. Civil 3D shows a preview of the vertical curve before you select the next PVI.

11   Press ENTER when finished. The layout profile is created with curves and labels.

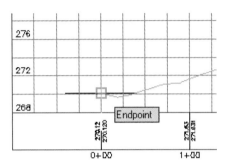

12   Zoom in to a section of the layout profile.

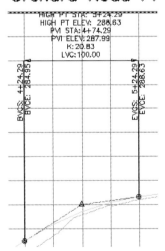

Notice that the tangent slopes are labeled as percent slopes. Also note that the vertical curve geometry points and data are labeled.

13 Close the Profile Layout Tools toolbar.

Next, you use a different method to create a second layout profile. First, you suppress the display of the Orchard Road - Design Profile.

14 In the drawing area, select the profile view.

15 On the ribbon, click Profile View Properties.

16 In the Profile View Properties dialog box, Profiles tab, for the Orchard Road - Design Profile, clear the Draw check box. Click OK.

The Orchard Road - Design Profile layout profile is removed from the profile view. You can always reverse this process.

17 On the command line, enter **–toolbar**. Press ENTER.

18 At the Toolbar Name prompt, enter **transparent_commands**. Press ENTER.

19 When prompted to enter an option, enter **S**. Press ENTER. The Transparent Commands toolbar is displayed.

20 On the ribbon, Home tab, Create Design panel, click Profile > Profile Creation Tools.

21 At the Select Profile View prompt, select the Orchard Road profile view.

22 In the Create Profile - Draw New dialog box, for Name, enter **Orchard Road - Design Profile (Opt 2)**. Click OK.

23 On the Profile Layout Tools toolbar, click Draw Tangents (the first button to the left on the Profile Layout Tools toolbar).

24 At the Specify Start Point prompt, using the Endpoint OSNAP, click the left end of the Existing Ground profile.

25 At the Specify End Point prompt, on the Transparent Commands toolbar, click Profile Station Elevation.

26 At the Select Profile View prompt, select the Orchard Road profile view.

27 At the Station prompt, enter **410 (145 m)**. Press ENTER.

28 At the Elevation prompt, enter **281.5 (86.5 m)**. Press ENTER.

29 Press ESC to cancel the current Transparent command.

30 On the Transparent Commands toolbar, click Profile Grade Length.

31 At the Grade prompt, enter **4.5**. Press ENTER.

32 At the Length prompt, enter **400 (120 m)**. Press ENTER.

33 Press ESC twice to cancel the current Transparent command and the Create
    Tangents command.

    Next, you create a vertical curve.

34 On the Profile Layout Tools toolbar, click Free Vertical Curve (Parameter).

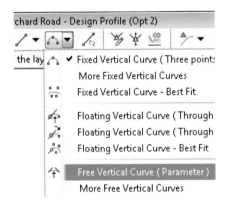

    Using this procedure, you click the tangent before the PVI, and then click the tangent
    after the PVI.

35 At the First Entity prompt, select the tangent to the left of the middle PVI.

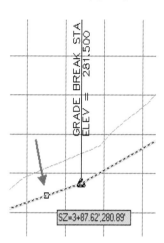

36 At the Next Entity prompt, select the tangent to the right of the middle PVI.

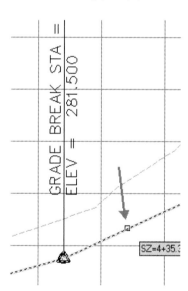

37   At the Curve Length prompt, enter **200 (60 m)**. Press ENTER twice.

A vertical curve is created between the two tangents.

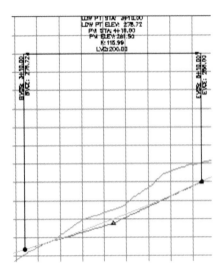

38   Close the Profile Layout Tools toolbar.

Next, you examine the profiles in Prospector.

39   In Toolspace, on the Prospector tab:

- Click to expand Alignments, Centerline Alignments, Orchard Road, and Profiles.

- Under Profiles, note the Orchard Road profile data.

40 Using grips, reposition the PVI labels at the beginning and end of the layout profile.

41 The completed drawing appears as follows.

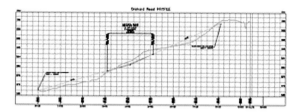

42 Close the drawing. Do not save the changes.

Lesson 25 | Editing Profile Geometry

This lesson describes how to edit the vertical geometry of layout profiles. You can edit layout profiles graphically, using the table in Panorama, or by using profile creation tools on the Profile Layout Tools toolbar. Profile entities are the individual components of a layout profile, specifically the tangents and the vertical curves.

Graphical editing of layout profiles involves using grips. When you use grips to edit layout profiles, vertical curves maintain tangency. You can also use tabular editing tools to manually change PVI elevations and stations, vertical curve lengths, K values, and grades. Profile layout tools enable you to change, delete, and add vertical alignment subentities. When you edit a profile, associated annotation is automatically updated.

Objectives

After completing this lesson, you will be able to:

- Use the available tools to edit profile geometry.

- List the guidelines for editing profile layout geometry.

- Edit a layout profile using graphical and tabular methods.

Tools for Editing Profile Geometry

You edit a layout profile graphically using tools such as grips and transparent commands. You can also make more precise edits to the layout profile geometry by changing data in a table such as grades, elevations, stations, and vertical curve design parameters.

The following illustration shows the grips that are displayed on a profile layout when you select it. You use the grips to move PVIs, move PVIs and maintain the entering or existing grade, change tangent locations, and change vertical curve lengths. When you move a PVI or tangent locations, connected vertical curves maintain their tangency.

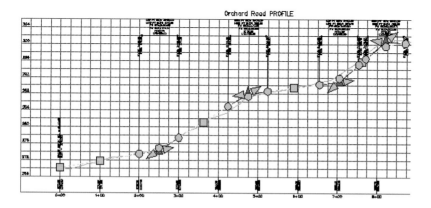

Graphical Editing

You use grips to graphically edit a layout profile. When you edit a layout profile with grips, you are editing the profile object. Therefore, any associated annotation and tabular data automatically updates. You click the layout profile to show the graphical editing grips.

Grips are displayed in the following illustration.

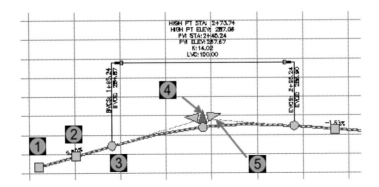

① Square grip at profile beginning and end tangents: change tangent grade.

② Square grip at tangent midpoints: move tangent and maintain grade.

③ Circular grips on vertical curves: resize vertical curves.

④ Red triangle grip on PVI: move the PVI.

⑤ Cyan triangle grips on PVI: move PVI and maintain grade of either incoming or outgoing tangent.

For more information, see "Profile Layout Tools" in Help.

Example of Using Grips to Move PVIs

The following illustrations show you how to use grips to change the location of a PVI. Click a grip to make it active, as shown in the following illustration.

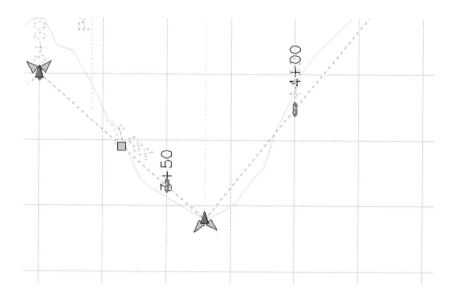

When the grip is active, you can move it to a new location, as shown. You can also select a new location more precisely using Object Snap or profile Transparent commands.

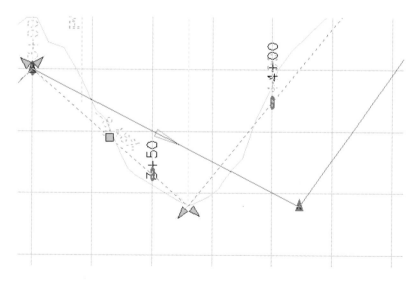

Grips are displayed until you cancel the selection of the profile. The following illustration shows the point at its new location.

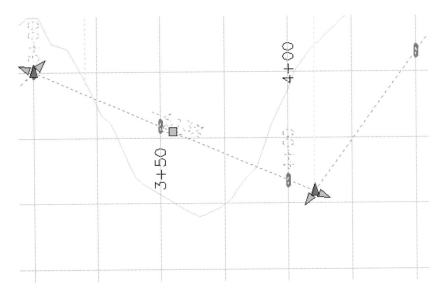

Editing with Layout Tools

You can also edit a layout profile by using commands on the Profile Layout Tools toolbar. In the drawing area, select the layout profile, right-click and click Edit Profile Geometry to show the Profile Layout Tools toolbar. You have a number of layout profile editing tools on the Profile Layout Tools toolbar. You can use any of the tangent and vertical curve commands to add profile entities. You can also use the Delete Entity button to remove profile entities. When you edit a profile with commands on the Profile Layout Tools toolbar, the graphical display of the profile and associated annotation automatically update.

From the Profile Layout Tools toolbar, you can also access the Panorama window to edit layout profile data in a table. You can view data in the table as either PVI-based or entity-based.

The Profile Geometry Editor command is available on the ribbon when the layout profile is selected. This is shown in the following illustration.

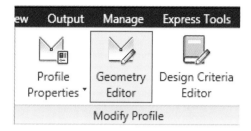

Guidelines for Editing Profile Layouts

Keep the following guidelines in mind when editing profile layouts.

- Layout profile grips work exactly the same as grips on all AutoCAD entities. Graphically editing a profile with grips is a three-step process. First click: select the layout profile. Second click: select the grip. Third click: reposition the grip.

- You can use transparent commands to help you to graphically edit the profile.

- Use the Tabular Editor in the Panorama window to round profile tangent grades, vertical curve lengths, and vertical curve K values.

Exercise 01 | Edit Profile Geometry

In this exercise, you edit a layout profile using graphical and tabular methods.

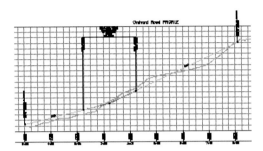

The completed exercise

1   Open \Site Design - Profiles\I_edit_profile_geometry.dwg (M_edit_profile_geometry.dwg).

2   In the profile view, zoom to the vertical curve.

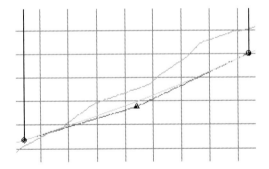

3   In the drawing area, select the layout profile. Review the different grips.

4   Click the middle PVI arrow grip. Move the mouse. Click again to move the PVI to the new location. Notice the behavior of the associated tangents.

5   On the Quick Access toolbar, click Undo.

6   Select the layout profile.

7   To the side of the PVI, click the left or right arrowhead grip. Move the mouse. Click to
     reposition the PVI.

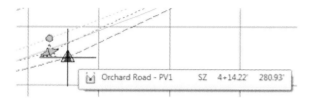

     The PVI position is moved directly along one of the tangents.

8   Click Undo.

9   Select the layout profile. On the ribbon, Modify Profile panel, click Geometry Editor.
     The Profile Layout Tools toolbar opens.

10   On the Profile Layout Tools toolbar, click Profile Grid View.

     The Profile Entities Panorama window opens, showing all of the geometric information about
     the profile.

11   Change the grade in values and round the grades to the nearest percent. Note that the layout
     profile dynamically updates after you make the change.

12   Click the arrowhead grip on the PVI. Move the mouse. Click to move the PVI.

13   In Panorama, notice the change in the value for the station and elevation of the PVI that was
     moved. Close Panorama.

14   Select the profile. Click the round grip at the beginning of the vertical curve. Move the
     mouse. Click to modify the curve length.

15   Select the profile. Click the square grip at the tangent midpoint. Move the mouse. Click to reposition the tangent.

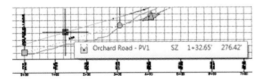

The PVI position can be modified using grips or by entering a station and elevation value in the Panorama window. You can also move, delete, and add PVIs using commands on the Profile Layout Tools toolbar.

16   On the Profile Layout Tools toolbar, click Insert PVI.

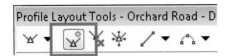

17   At the Specify Point for New PVI prompt, click a location in the profile view for a new PVI. Press ENTER.

Notice that a curve is not automatically positioned, but a grade break label is inserted.

18   On the Profile Layout Tools toolbar, click Delete PVI.

19   At the Pick Point Near PVI to Delete point prompt, click near the new PVI. Press ENTER.

Notice that the PVI and label are removed.

20   On the Profile Layout Tools toolbar, click Insert PVIs - Tabular.

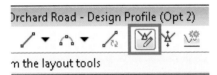

21 In the Insert PVIs dialog box:

- Under Vertical Curve Type, click Parabolic.

- Notice the Curve Length box is added. You can add the PVI and curve parameters directly in this box.

- Click Cancel.

22 On the Profile Layout Tools toolbar, click Raise/ Lower PVIs.

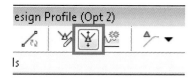

23 In the Raise/Lower PVI Elevation dialog box:

- For Elevation Change, enter **0.5' (0.2 m)**.

You can raise or lower the entire profile or just a station range.

- Click OK.

The entire profile is raised 0.5 ft (or 0.2 m).

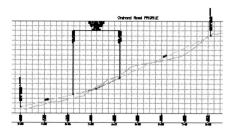

24 Close the drawing. Do not save the changes.

Lesson 26 | Labeling Profiles and Profile Views

This lesson describes how to label profiles and profile views. Much of the labeling on a profile view is generated directly from the profile view style. The profile view style displays labeling on both axes and on the profile view title. You can also add station/elevation and depth labels to the profile view as an independent operation.

You label surface profile data after the surface profile has been created. Layout profiles are labeled when the profile is created, or after the profile has been created. You can label profiles with stations, tangent grades, vertical curve data, and grade breaks. You can also label horizontal alignment geometry on the profile.

Profile label sets contain the individual profile label types. All profile view and profile labels automatically update when the profile data changes.

Objectives

After completing this lesson, you will be able to:

- Describe profile and profile view labels, and label sets.

- Explain how you edit profile labels.

- Use guidelines to label profiles and profile views.

- Label profiles and profile views.

About Profile and Profile View Labels

When you create a layout profile, you can automatically label it using label styles. Profile label styles can be configured to mark any of the following standard points along the profile:

- Major and minor stations of the parent horizontal alignment.

- Horizontal geometry points.

- Grade breaklines.

- Sag curves.

- Crest curves.

Profile label sets are the collection of individual profile label types. On the Settings tab of the Toolspace window, you can find the profile label sets by expanding the Profile, Label Styles, and Label Sets trees.

Definition of Profile Labels

You can label layout profiles with tangent grades, vertical curve data, grade break data, alignment stationing, and horizontal alignment geometry label types.

Individual label types can be applied to a layout profile. Alternatively, if the layout profile has already been created, you can add profile labels. A common method is to combine individual label types into a label set and apply the label set to the layout profile. This enables you to standardize on collections of labels that you would regularly apply to profiles. When you create a layout profile, you can choose a label set to apply to the profile.

When you create a layout profile, you choose a label set to apply to the profile, as shown in the following illustration.

Definition of Label Sets

Label sets are collections of profile labels. You combine vertical curve, tangent, grade break, station, and horizontal alignment geometry labels in a label set. When you create a layout profile, you can apply the label set to the profile. The label set can also be applied to the profile afterward.

Profile View Labels

The profile view style accounts for much of the labeling on a profile view. With the profile view style, you can label the horizontal axes, the vertical axes, and the title. Once you change a profile view style, all profiles view that reference that style automatically update to reflect the new display parameters. Use the Display tab on the Profile View Style dialog box to determine which labeling components of the profile view are displayed, as shown in the following illustration.

Component Type	Visible	Layer
Graph Title	♀	C-ROAD-PROF-TITL
Left Axis	♀	C-ROAD-PROF-TTLB
Left Axis Title	♀	C-ROAD-PROF-TITL
Left Axis Annotation Major	♀	C-ROAD-PROF-TEXT
Left Axis Annotation Minor	♀	C-ROAD-PROF-TEXT
Left Axis Ticks Major	♀	C-ROAD-PROF-TEXT
Left Axis Ticks Minor	♀	C-ROAD-PROF-TEXT
Right Axis	♀	C-ROAD-PROF-TTLB
Right Axis Title	♀	C-ROAD-PROF-TITL

Example

When you apply profile labels to a profile, the default placement of the profile labels is determined in the profile label set. In the Profile Label Sets dialog box you can specify the following:

- Types of labels to be added to the label set, which include sag curves, crest curves, tangents, major and minor stations, grade breaks, and horizontal geometry points.

- Style of each type of label to be included in the label set.

- Starting and ending stations for label placement.

- Default placement options for the labels.

Part of the Profile Label Set dialog box is displayed in the following illustration.

Editing Profile Labels

You can change the position of the profile labels after they are created. You either edit the position of the labels graphically using grips, or you edit the position of the profile labels in a table. The latter option displays a dialog box similar to the Profile Label Set dialog box, where you change the Dimension Anchor option and Dimension Anchor value in a table.

When you edit profile labels, you apply a label set or apply individual labels to the profile.

Editing Profile Labels

You can use the following Dimension Anchor options:

- Distance Above and Below PVI.

- Fixed (Absolute) Elevation.

- Distance Relative to the Profile View Top and Profile View Bottom.

When you use grips to modify the position of the labels, drag the labels to their new locations. The Label Dimension Anchor grip enables you to move the entire vertical curve up and down. The square grip is used to move the profile label so that it adopts its dragged state display property, as shown in the following illustration.

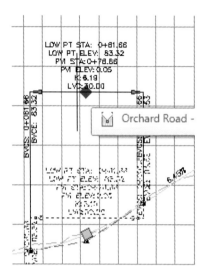

Guidelines for Labeling Profiles and Profile Views

Keep the following guidelines in mind when creating profile labels and label sets.

- You can create a profile label set called None, which can be applied to a layout profile when you first create it. This action makes it easier to visualize your design geometry.

- Profile styles and label styles should be created and stored in your drawing template. All new drawings created with the drawing template carry the styles forward.

- You can set the default profile view, and profile styles and label styles, by modifying the command settings.

Exercise 01 | Label Profiles and Profile Views

In this exercise, you add labels to profiles and profile views. You also modify label styles for both objects.

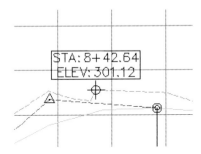

The completed exercise

1   Open...*Site Design - Profiles\I_label_profiles_profileviews.dwg* (*M_label_profiles_profileviews.dwg*).

2   In Toolspace, Settings tab:

- Expand Profile, Label Styles, Label Sets.

- Right-click Complete Label Set, click Edit.

3   In the Profile Label Set - Complete Label Set dialog box, Labels tab:

- Review the four types of labels included in this label set: Lines, Grade Breaks, Sag Curves, and Crest Curves.

- For Lines, in the Style column, click the Style icon.

- Select Slope 1 in X from the list.

- Click OK twice to close all dialog boxes.

This change affects the Complete Label Set only. To see changes in the profile labels, you need to import the label set for that profile. Another option is to modify the profile labels in use directly.

4   In the drawing area, select the layout profile.

Right-click, click Edit Labels.

5   In the Profile Labels - Orchard Road - Design Profile dialog box:

- Click Import Label Set.

- Select Complete Label Set from the list.

- Click OK twice to close all dialog boxes.

6   Zoom in to a tangent on the profile and view the change in the label style.

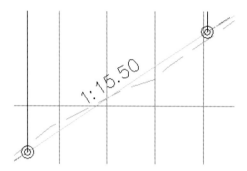

Each profile can be modified independent of the label set.

7   Select the layout profile. Right-click, click Edit Labels.

8   In the Profile Labels - Orchard Road - Design Profile dialog box:

- For Lines, click the Style icon.

- Select Percent Grade from the list.

- Click OK twice to close both dialog boxes.

9   Zoom in to a tangent grade label to view the different style.

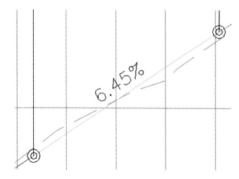

10   In the drawing area, review the labeling for the vertical curves.

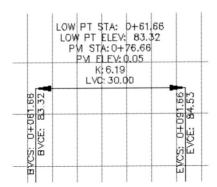

BVCS is the Beginning Vertical Curve Station. BVCE is the Beginning Vertical Curve Elevation. Similarly, EVCS and EVCE are the same values for the end of the curve. Low point and PVI station and elevation are included, as well as K value and the LVC (Length of Vertical Curve).

11   In Settings:

- Expand Profile, Label Styles, Curve, Crest and Sag.

- Right-click Crest Only. Click Edit.

12   In the Label Style Composer - Crest Only dialog box, Layout tab.

- For Component, click PVI Sta and Elev.

- Under Text, Color, click in the Value cell. Click Color Palette.

- In the palette, click red (color 1). Click OK.

- Notice the change in the Preview box to the right.

- Click OK to close the dialog box.

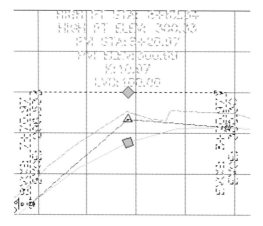

HIGH PT STA: 5+24.29
HIGH PT ELEV: 288.63
PVI STA: 4+74.29
PVI ELEV: 287.99
K: 20.83
LVC: 100.00

BVCS: 4+24.29
BVCE: 284.95
EVCS: 5+24.29
EVCE: 288.63

13   Zoom in to a crest curve.

14   On the profile:

- Select a vertical curve label.

- Click the diamond grip on the dimension line. Move the entire label up and down within the profile view.

- Click the square grip on the curve.

Reposition the label to show its dragged state display property. Next, you examine the profile view labels.

15   In the drawing area, select the Orchard Road profile view.

16 On the contextual ribbon, click Add View Labels > Station Elevation.

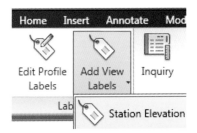

17 When prompted to:

- Select the profile view, click the profile view grid.

- Specify station, move the mouse left or right, then click to set the desired station for the label.

- Specify elevation, move the mouse up and down, click to set the elevation for the desired point to be labeled.

- Press ESC to end the command.

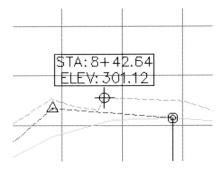

18 Close the drawing. Do not save the changes.

Chapter 08

Site Design - Assemblies and Corridors

In this chapter, you create corridor models for subdivision roads. Corridor models for large subdivisions are often calculated to the limits of the right of ways, with no daylighting. The grading for the adjacent land parcels is created from the 3D corridor feature lines at the right of way locations.

First, you create a symmetrical assembly for modeling the tangent section of the subdivision road, and an asymmetrical assembly for modeling the cul-de-sac. The assemblies are created by adding pre-defined subassemblies for the lane, curb and sidewalk cross section elements to the assembly objects.

Next, you create a corridor model from two baselines, or alignments. The first baseline is for the centerline alignment for the tangent section of the road and references the symmetrical assembly. The second baseline for the road is the edge of pavement alignment for the cul-de-sac and references the asymmetrical assembly.

Finally, you create top and datum corridor surfaces. The corridor top surface can be used for labeling design spot elevations and grades, and can also be used to calculate pipe network rim and invert elevations. The corridor datum surface can be used to calculate earth cut and fill quantities.

▶ Objectives

After completing this chapter, you will be able to:

- Create assemblies and add subassemblies.

- Create a simple model of a residential subdivision road corridor.

- Create corridor surfaces from corridor models.

Lesson 27 | Creating Assemblies

This lesson describes how to create assemblies and add subassemblies to an assembly.

The assembly object, along with the horizontal and vertical alignment, is used to build the corridor model for a road, highway, railway, embankment, channel, or any cross section-based features. The subassemblies are logically designed, and respond dynamically in the design environment, making it easy to generate and evaluate design alternatives.

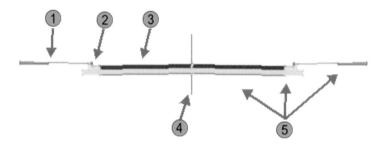

① Sidewalk

② Curb

③ Lane

④ Assembly

⑤ Subassemblies

Objectives

After completing this lesson, you will be able to:

- Describe assemblies and subassemblies.
- Describe the subassembly input and target parameters.
- Explain how to use subassemblies to build assemblies.
- Create assemblies.

About Assemblies and Subassemblies

Assemblies represent the typical section and are the starting point for corridor design. To create the assembly, you add subassemblies for the lanes, curbs, shoulders, and other cross-section elements.

When you create an assembly, you can make it available for future projects and other users by saving it on a tool palette, or within the drawing template (DWT) file. You can also save collections of assemblies in an assembly set.

Definition of Assemblies

An assembly is an arrangement of cross-section features found on a roadway or other corridor. It represents a typical section of the corridor that you position with an alignment and a profile. You create an assembly and then add subassemblies for cross-section elements such as lanes, curbs, sidewalks, shoulders, and side slopes.

Definition of Subassemblies

A subassembly is the basic building block that makes up an assembly. A subassembly is attached to one or both sides of the assembly's baseline, and subsequent subassemblies are attached to the appropriate points of the previously attached subassemblies. An assembly can be defined by attaching all of the subassemblies to one side of a baseline, and then mirroring these subassemblies to the other side of the baseline.

Subassemblies are intelligent objects that dynamically react to changes in the design environment. Each subassembly has its own set of parameters that you can modify to change its appearance or behavior. Civil 3D® provides a library of the most common, generic subassemblies that you may encounter in roadway design.

Along with using the predefined subassemblies in Civil 3D, you can also draw your own subassemblies. Civil 3D has the tools to convert your polyline shape into a subassembly. This subassembly has limited logic but can be swept down an alignment like any other subassembly. You can build your own custom subassembly with all of the parameters and functionality (or more) of those supplied in the subassembly catalog.

Assembly Elements

An assembly is made up of the following elements.

Element	Description
Baseline	The vertical line used as a display reference line for the assembly.
Baseline point	The point to which you attach subassemblies, and the point on the assembly attached to the horizontal and vertical alignment to create the corridor model. Also known as horizontal and vertical control.
Subassemblies	Cross-section element objects such as lanes, curbs, shoulders, and side slopes that you add to the assembly object. The subassemblies are added from the tool palette and attached to the assembly baseline or other subassemblies in the assembly.

Assembly Example

The following illustration shows a simple assembly. On either side of the baseline are subassembly objects that represent a lane, shoulder, guardrail, and cut or fill slope. In this example, the subassemblies are arranged starting from the baseline point, which is indicated by the circle with a square inside it at the midpoint of the baseline.

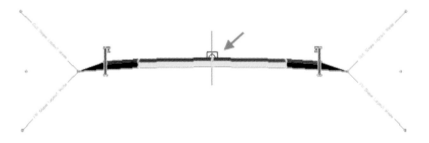

Subassembly Parameters

Subassemblies have input and target parameters that are used to change their geometric configuration.

Subassembly Input Parameters

Subassembly input parameters control the size, shape, and geometry of the subassembly. Custom subassembly input parameters can be saved on a tool palette and also specified when you add the subassembly to an assembly.

For example, the most common subassembly for modeling a lane is the LaneOutsideSuper subassembly. The LaneOutsideSuper subassembly has input parameters that set the general configuration values for the lane such as width, pavement depth, and cross fall. The default input parameters for the LaneOutsideSuper subassembly are shown in the following illustration.

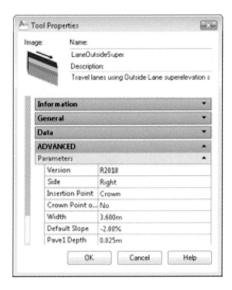

Target Parameters

Some subassemblies have required or optional target parameters. Target parameters control how the subassembly functions. Function is determined by the object the subassembly connects to (alignment, profile, or surface).

The daylighting subassemblies have required surface target parameters. You are required to specify a target surface to which the subassembly daylights.

The lane subassemblies have optional width and elevation target parameters that can be alignments and profiles. You can optionally override the lane width input parameter by targeting an alignment. This is how you vary the width of the lane in a lane taper. You can also optionally override the default slope input parameter by targeting a profile to change the lane cross fall.

Examples

For daylighting subassemblies that are used to project match slopes to surfaces, you are required to specify a surface model as the target parameter. Without the target parameter, the subassembly would not function. Lane subassemblies have optional target parameters such as lane width. A default input parameter for a lane subassembly is the lane width. The optional lane width target parameter would use another alignment to override the default lane width value. This is useful when pavement widths vary at turning lanes and lane tapers.

The most common subassembly to daylight to a surface is called BasicSideSlopeCutDitch. The BasicSideSlopeCutDitch subassembly has a required target parameter for a surface. When you create the corridor model with an assembly that contains BasicSideSlopeCutDitch, you must specify the target surface for the daylighting subassembly. This is shown in the following illustration.

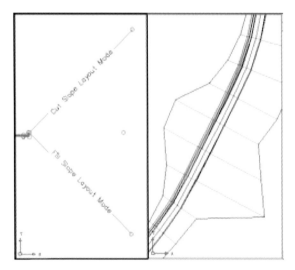

The image on the left shows the subassembly as part of the right side of an assembly, attached to a curb and gutter. The image on the right shows the result of the logic built into the subassembly when it is used in a corridor model. The width of the slope adjusts to account for changes in the terrain.

Creating Assemblies

This section describes the process for creating assemblies. You create an assembly by adding the assembly to the drawing area, and then using the baseline and baseline point as a visual guide for the addition of subassemblies. Sets of subassemblies are included in the tool palettes. You can also select subassemblies from the subassembly catalog and place them on the tool palettes.

The following illustration shows a basic assembly with baseline, basic lane, and curb and gutter subassemblies.

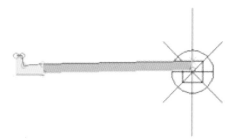

Process Description

To create an assembly you launch the Create Assembly command from the ribbon, assign a name to the assembly, and insert the assembly to the drawing. The assembly object is shown in the following illustration.

After you insert the assembly object you begin by adding the subassemblies to the assembly. The first subassemblies you add are added to the assembly and you can pick any part of the assembly object. As you move away from the assembly you need to add subassemblies to other subassemblies. You select the subassembly markers as insertion points for the other subassemblies. In the following illustration, the arrow points to one of subassembly markers.

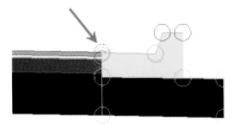

Guidelines

Keep the following guidelines in mind when you create assemblies.

- For every subassembly that you add to your assembly, you should read the Help file to understand how the subassembly behaves and which default settings you can modify.

- Modify the default subassembly input parameters on the tool palette to suit common design configurations.

- Rename the subassemblies on the tool palette to clearly indicate key geometric properties, such as width, slope, and grade.

- Subassemblies added to the assembly are organized into assembly groups. When you select the assembly object to add subassemblies, a new assembly group is created. When you modify the properties of the assembly, rename the groups with indicative names, such as LEFT, RIGHT, and so on.

- To create a mirror image of a subassembly or group of subassemblies, use the Mirror command, which is available on the shortcut menu.

- Once the assembly is complete, you should go back and rename each of the subassemblies to give them descriptive names. This may not seem important early in a project when you have a simple assembly, but roadway jobs can get very complex. Numerous assemblies can be applied to any one alignment. It is good practice to document the components of an assembly so that other team members working on the project understand the parts and their functions.

Exercise 01 | Create Assemblies

A typical section shows the engineering details of a cross-section configuration. Typical sections are represented with assemblies in Civil 3D.

In this exercise, you create two assemblies for the Cedar Cove cul-de-sac corridor. One is required for the tangent section, and the other is required for the bulb at the end of the cul-de-sac.

The completed exercise

1   Open *Site Design - Assemblies and Corridors\I_create_assembly.dwg* (*M_create_assembly.dwg*).

     Start by configuring your workspace. To create assemblies, you need to view the AutoCAD Object Properties window with the tool palettes.

2   Close Toolspace.

3   On the command line, enter **PR** (Object Properties). Press ENTER.

4   Dock the Properties window on the left side of the screen. Right-click the title bar. Toggle Auto- Hide off.

5   On the command line, enter **TP** (Tool Palettes). Press ENTER.

6   Dock the Tool Palettes window on the right side of the screen. Right-click the title bar. Toggle Auto-Hide off.

     The Properties window and the Tool Palettes window are both expanded and docked on either side of the screen.

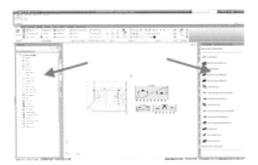

Next, you configure the Tool Palettes window to show subassemblies in your unit of measure: imperial or metric.

7   Right-click the Tool Palettes title bar, click Civil Imperial Subassemblies (Civil Metric Subassemblies).

Next, you create the assembly for the tangent section of the cul-de-sac. This is a symmetrical assembly that models both sides of the road.

8   On the ribbon, Home tab, Create Design panel, click Assembly > Create Assembly.

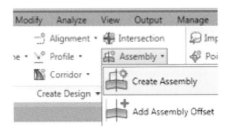

9   In the Create Assembly dialog box, for Name, enter **2 Lane Urban**. Click OK.

10   At the Specify Assembly Baseline Location prompt, in the drawing, select a location to create the assembly.

The assembly object is inserted and zoomed to in the drawing.

Next, you add the lane subassemblies on the right and left sides.

11   On the tool palettes, click the Lanes palette.

12   Click LaneOutsideSuper.

13   In the Properties dialog box:

- Click to collapse the Information, General, and Data sections.

- Under Advanced Parameters, for Side, click Right.

- For Width, enter **16' (4m)**.

- Press ENTER.

14   At the Select a Marker Point within Assembly prompt, click the assembly.

The subassembly is attached to the assembly marker point. You see the lane structure for the right side in the drawing.

Next, you insert the left lane.

15   In the Properties dialog box, for Side, click Left.

16   At the Select a Marker Point with Assembly prompt, click the assembly in the same location to insert the left lane.

Next, you insert the curb and gutter subassemblies.

17   On the tool palettes, click the Curbs palette.

18   Click UrbanCurbGutterGeneral.

19   In the Properties dialog box, for Side, click Right.

20   At the Select a Marker Point within Assembly prompt, click the subassembly marker point at the top of pavement edge location on the right side.

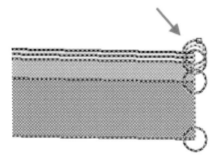

The UrbanCurbGutterGeneral subassembly is inserted on the right side.

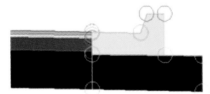

21   In the Properties window, for Side, click Left.

22   At the Select a Marker Point within Assembly prompt, click the subassembly marker point at the top of pavement edge location on the left side.

      Note: If you make a mistake, you can use the AutoCAD Erase command to erase the subassembly and try again.

      Next, you insert the sidewalks.

23   On the Curbs palette, click UrbanSidewalk.

24   In the Properties dialog box:

      • For Side, click Right.

      • For Inside Boulevard Width, enter **7' (2.4m)**.

      • For Sidewalk Width, enter **4' (1m)**.

      • For Outside Boulevard Width, enter **1' (.5m)**.

25   At the Select a Marker Point within Assembly prompt, click the subassembly marker point on the top and back of curb on the right side.

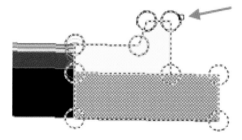

      The sidewalk is added to the assembly.

26 In the Properties dialog box, for Side, click Left.

27 At the Select a Marker Point within Assembly prompt, click the subassembly marker point on the top and back of curb on the left side.

You now have sidewalks on both sides of the assembly. The completed assembly appears as follows.

Next, you create the assembly for the bulb section of the cul-de-sac.

28 On the ribbon, Home tab, Create Design panel, click Assembly > Create Assembly.

29 In the Create Assembly dialog box, for Name, enter **Half Section**. Click OK.

30 At the Specify Assembly Baseline Location prompt, click below the 2 Lane Urban assembly.

31 On the Curbs palette, click UrbanCurbGutterGeneral.

32 In the Properties dialog box, for Side, click Right.

33 At the Select Marker Point within Assembly prompt, click the assembly baseline to insert the subassembly.

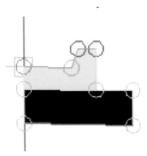

Chapter 08 | Site Design - Assemblies and Corridors

34 On the Curb palette, click UrbanSidewalk.

35 In the Properties dialog box:

- For Side, click Right.

- For Inside Boulevard Width, enter **7' (2.4m)**.

- For Sidewalk Width, enter **4' (1m)**.

- For Outside Boulevard Width, enter **1' (0.5m)**.

36 At the Select Marker Point within Assembly prompt, click the subassembly marker on the top of the back of the curb.

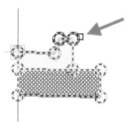

The sidewalk is inserted.

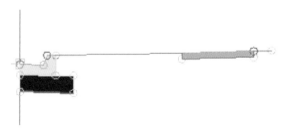

Next, you add a lane on the left side of the assembly. This lane forms the center paved portion of the cul-de-sac when the assembly is processed around the pavement edge baseline in the cul-de-sac.

37 Click the Lanes palette.

38 Click LaneOutsideSuper.

39 In the Properties dialog box:

- For Side, click Left.

- For Width, enter **16' (4m)**.

40 At the Select Marker Point within Assembly prompt, click the assembly. The lane subassembly is inserted.

The Half Section assembly appears as shown.

41 Press ESC to cancel the command.

42 Close the Properties dialog box.

The completed exercise appears as follows.

43 Close the Tool Palettes and Properties windows.

- Open the Toolspace window.

44 Close the drawing. Do not save the changes.

Lesson 28 | Creating Corridor Models

This lesson describes the components of a corridor model, and how to use them to build a simple model of a residential subdivision road with a cul-de-sac.

You can use corridor models to represent any road, rail, channel, or berm design that has typical cross- section features. When you create a corridor model, you create a single object that includes all the design components and input parameters for a road or other type of feature created from a typical cross-section.

A completed corridor model for a residential subdivision road with a cul-de-sac is shown in the following illustration.

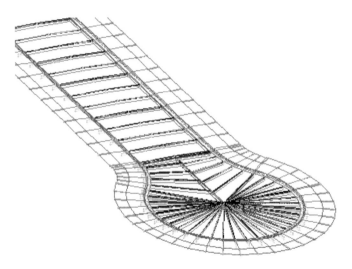

Objectives

After completing this lesson, you will be able to:

- Describe a corridor model and list its components.

- Explain the process for creating a subdivision corridor model.

- Create a corridor model for a subdivision road with a cul-de-sac.

About Corridor Models

When you create a corridor, you can create complex models for transportation facilities such as roads, highways, and railways by adding corridor regions or referencing more than one alignment baseline. Almost any condition that your design may call for can be modeled by using the corridor functionality.

A simple corridor creates a model based on a single alignment (baseline), profile, and assembly. You can make corridor models more complex by adding additional alignments (baselines), or by creating regions that use different assemblies. You can enhance and expand the basic corridor configuration to accommodate a wide variety of designs.

Definition of Corridor Models

A corridor model is a three-dimensional representation of the design for a roadway, railway, or other transportation facility. Corridors can also be used to model channels, berms, and any other civil engineering feature that can be represented with a typical section. You create a corridor model with an alignment, a layout profile, and an assembly. By incorporating constraints and rules into the corridor design, you can manage and control how the corridor model interacts with the terrain and other alignments and profiles.

Definition of Subdivision Corridor Models

Many large subdivision designs are created based on developing an interim grading surface that accounts for site topography and drainage constraints. The subdivision roads are usually the spine of the subdivision and designed first. The layout profiles are designed to match surface profiles generated from an interim grading surface.

Unlike road and highway corridor models, the assemblies used to create subdivision corridor models typically do not include daylighting subassemblies and therefore do not interact with surface models. Subdivision corridor models are usually calculated up to the right-of-way and are essentially floated through space along the horizontal and vertical alignments.

A typical assembly for a subdivision corridor model is shown in the following illustration. The right-of- ways are indicated by the arrows.

Additional Components

In addition to the alignment, layout profile, and the assembly, corridors are also comprised of feature lines, regions, and baselines.

Feature Lines

Feature lines are the longitudinal lines along the corridor that are used to represent typical cross-section points such as road crowns, pavement edges, gutter flow lines, sidewalk edges, and daylighting lines. Feature lines are part of the corridor model and are created by connecting the subassembly points of the corridor assembly.

Regions

Corridor regions are independent station ranges to which you can apply different assemblies to model differing cross-section configurations.

Baseline

The baseline is the alignment that is used to control the creation of a corridor model. A corridor model can be constructed from several baselines. A corridor for a subdivision road with a cul-de-sac uses a centerline alignment baseline for the main section of the road, and an edge of pavement alignment baseline for the cul-de-sac.

The following illustration shows feature lines on a corridor model for a subdivision road with a cul-de- sac.

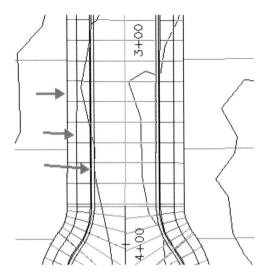

Corridor Examples

A corridor model of a cul-de-sac is shown in the following illustration.

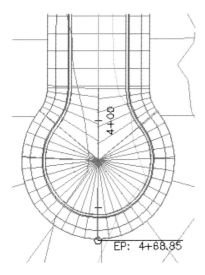

Complex Corridor Examples

The following illustration shows three centerline alignments that represent an on-ramp (entering from the top- right corner), an off-ramp, and the two-lane road where the two ramps join. The alignments for the on- and off-ramps end at precisely the same location as the beginning of the two-lane segment, which creates a continuous corridor from the three segments. The illustration shows the same alignments after a corridor is created. Different assemblies are applied along the alignments to create a single corridor with lanes and ditches that merge together seamlessly. Each alignment is associated with the corridor model by using a different baseline entry in the corridor properties. The illustration shows a transition between regions in a corridor. The upper region of the corridor uses an assembly with a plain shoulder. The lower region uses a similar assembly that has a curb and gutter instead of the paved shoulder.

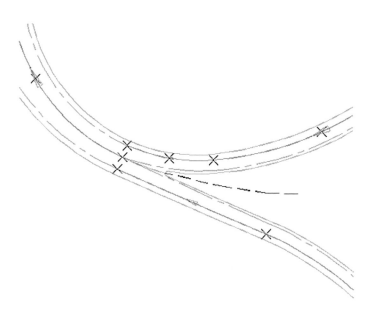

The following illustration shows the same alignments after a corridor is created. Different assemblies are applied along the alignments to create a single corridor with lanes and ditches that merge together seamlessly. Each alignment is associated with the corridor model by using a different baseline entry in the corridor properties.

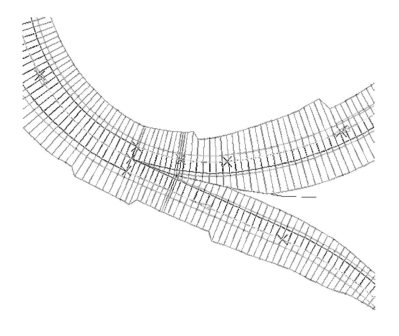

The following illustration shows a transition between regions in a corridor. The upper region of the corridor uses an assembly with a plain shoulder. The lower region uses a similar assembly that has a curb and gutter instead of the paved shoulder.

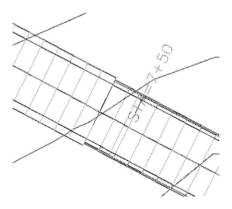

Creating Corridor Models

When you create a corridor, it is displayed on the Prospector tool under the Corridors collection. You can edit the corridor later to add more complex features such as baselines and regions.

Creating Corridor Models

To create a corridor model, you must specify an alignment, a vertical alignment, and an assembly. The assembly is processed along the horizontal alignment at a prespecified "assembly insertion frequency," which is the increment along the alignment where the assembly is inserted to create the corridor. This is shown in the following illustration.

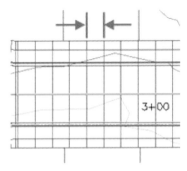

The assembly insertion frequency is not related to the increment at which cross-sections can be created, and can be changed when you modify the properties of the corridor.

The alignment you specify when you create the corridor is called a baseline. The baseline is the alignment that is used to control the creation of a corridor model. You can specify multiple baselines when creating a corridor. This is how you create a corridor that incorporates a cul-de-sac. Complex corridor models can incorporate several baselines.

If the assembly you use has subassemblies that require target assignments, such as the daylight subassemblies, when you create the corridor you have the option to specify the targets. The target for a daylight subassembly would be a surface model. Typically, subdivision assemblies do not use daylight subassemblies.

Corridor regions are segments of a corridor that are defined by an independent station range and assembly. When you use multiple corridor regions, you can use multiple assemblies to model the corridor over a different station range. This makes it easy to model widely variable cross-section configurations with a single corridor model.

Creating Corridors with a Cul-de-Sac

A corridor model for a subdivision road with a cul-de-sac incorporates two baselines. The first baseline controls the creation of the corridor for the main part of the road and is usually the road centerline alignment. The second baseline controls the creation of the corridor cul-de-sac and is usually the edge of pavement or gutter alignment around the cul-de-sac. This is shown in the following illustration.

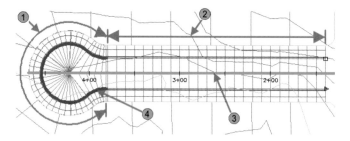

1. Second Baseline Region 1

2. First Baseline Region 1

3. First Baseline

4. Second Baseline

Creating Corridors with a Cul-de-Sac

The assembly that you process along the first baseline represents a symmetrical cross-section up to the right-of-way locations. The assembly you process along the second baseline is an asymmetrical cross-section with the baseline at the edge of pavement location. The assembly baseline is indicated by the arrow in the following illustration.

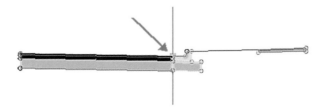

The lane subassembly in the cul-de-sac assembly has a fixed width. The width of the lane subassembly and the elevation of the inside crown is overridden by assigning the optional alignment and profile target to the centerline alignment and profile of the cul-de-sac alignment. This essentially stretches the lane back to the centerline alignment and elevation in the cul-de-sac. The following illustration shows the completed cul-de-sac without targets assigned to the lane subassembly.

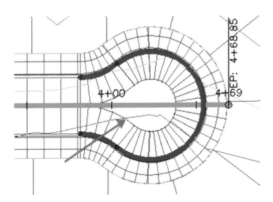

When you assign targets to the lane subassembly on the asymmetrical assembly for the second corridor region, the lane stretches back to the centerline alignment location and elevation. This is shown in the following illustration.

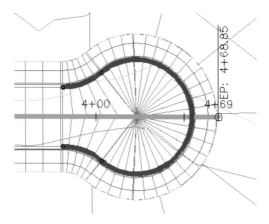

Guidelines

Keep the following guidelines in mind when you create corridor models.

- When you create a corridor model, you can use the Create Corridor command or the Create Simple Corridor command. Both commands result in the same object. However, the Create Simple Corridor commands offers fewer input options to create the corridor.

- When you create the corridor model, make sure that you specify a station range for each region so that you can assign the appropriate target. Ensure that you have layout profile data for the station range over which the corridor regions are created.

- Use representative naming conventions for alignments, profiles, corridors, and corridor regions. This makes it easier to create and manage corridor data.

Exercise 01 | Create a Corridor Model

In this exercise, you create a corridor model for a subdivision road with a cul-de-sac.

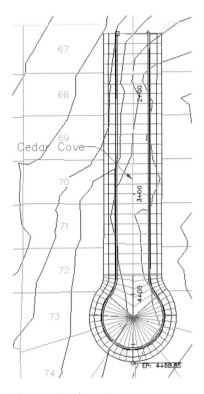

The completed exercise

1   Open \Site Design – Assemblies and Corridors\I_create_corridor.dwg (M_create_corridor.dwg).

2   Zoom in to the Cedar Cove alignment on the east side of the site.

You begin by modeling the tangent section of the corridor.

3   On the ribbon, Home tab, Create Design panel, click Corridor > Create Corridor.

4   At the Select a Baseline Alignment prompt, in the drawing area, click the Cedar Cove centerline alignment.

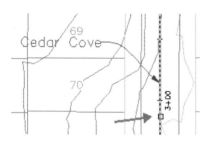

5   At the Select a Profile prompt, press ENTER.

6   In the Select a Profile dialog box, click Cedar Cove - FG. Click OK.

7   At the Select an Assembly prompt, press ENTER.

8   In the Select an Assembly dialog box, click 2 Lane Urban. Click OK.

9   In the Create Corridor dialog box:

    - For Corridor Name, enter **Cedar Cove**.

    - In the Name column, click RG - 2 Lane Urban - (1). Change the name to **Tangent**.

    - In the Start Station column, click the Select Station.

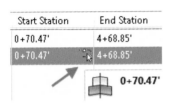

10 At the Specify Station along Alignment prompt, in the drawing area, snap to the endpoint of
    the edge of the pavement alignment to the right of Parcel 67.

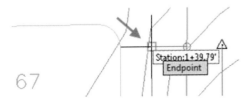

The Tangent region of the corridor starts at this location.

11   In the Create Corridor Dialog box, for Tangent, in the End Station column, click the Select Station.

12   At the Specify Station along Alignment prompt, snap to the endpoint of the first arc at the northwest end of the cul-de-sac.

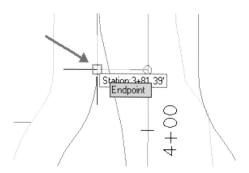

The Tangent region of the corridor ends at this location.

13   In the Create Corridor Dialog box, for Tangent, in the Frequency column, click the ellipse.

14   In the Frequency to Apply Assemblies dialog box, under Apply Assembly:

- For Along Tangents, enter **10 (5 m)**.

- For Along Curves, enter **10 (5 m)**.

- For Along Spirals, enter **10 (5 m)**.

- For Along Profile Curves, enter **10 (5 m)**.

The assembly will be processed every 10' (5 m) along this region to calculate the corridor.

15   Click OK twice to close the dialog boxes.

The corridor for the tangent region of Cedar Cove is created.

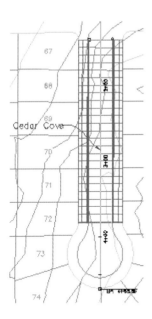

Next, you model the cul-de-sac.

16 In the drawing area, select the corridor. On the ribbon, click Corridor Properties.

17 In the Corridor Properties - Cedar Cove dialog box, Parameters tab, click Add Baseline.

18 In the Create Corridor Baseline dialog box, for Horizontal Alignment, click Cedar Cove EP. Click OK.

19 For Cedar Cove EP - (2):

- Click in the Profile column.

- In the Select a Profile dialog box, click Cedar Cove EP - FG.

- Click OK.

20 Right-click Baseline - Cedar Cove EP – (2). Click Add Region.

21 In the Create Corridor Region dialog box, for Assembly, select Half Section assembly from the list. Click OK.

22 Expand Baseline - Cedar Cove EP – (2).

23 Double-click RG – Half Section – (1). Change the name to Cul-de-sac.

24 For the Cul-de-sac region, under Start Station, click Select Station.

25 At the Specify Station along Alignment prompt, snap to the endpoint of the first arc on the cul- de-sac.

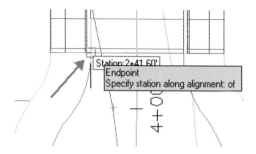

26 For the Cul-de-sac region, click Select Station in the End Station column.

27 At the Specify Station Along Alignment prompt, snap to the endpoint of the last arc on the cul- de-sac.

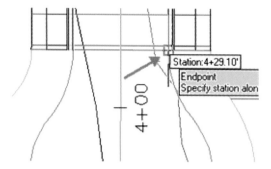

28   For the Cul-de-sac region, under Frequency, click the ellipse.

29   In the Frequency to Apply Assemblies dialog box, under Apply Assembly:

- For Along Tangents, enter **5 (2 m)**.

- For Along Curves, enter **5 (2 m)**.

- For Along Spirals, enter **5 (2 m)**.

- For Along Profile Curves, enter **5 (2 m)**.

- Click OK twice to close the dialog boxes.

The corridor is constructed into the cul-de-sac. Notice that the cul-de-sac does not extend back to the center of Cedar Cove. The lane width for the half section is too small.

30   In the drawing area, select the corridor. Right- click, click Corridor Properties.

31   In the Corridor Properties – Cedar Cove dialog box, Parameters tab, for the Cul-de-sac region, under Target, click the ellipse.

You use target mapping to stretch the LaneOutsideSuper assembly to the Cedar Cove centerline and profile. This completes the design to the center of the cul-de-sac.

32   In the Target Mapping dialog box, for Width Alignment, click in the Object Name cell.

33   In the Set Width or Offset Target dialog box:

- For Select Object Type to Target, click Alignments.

- Under Select Alignments, click Cedar Cove.

- Click Add.

- Click OK.

34   For Outside Elevation Profile, click in the Object Name cell.

35   In the Set Slope or Elevation Target dialog box:

- For Select Object Type to Target, click Profiles.

- For Select an Alignment, click Cedar Cove.

- For Select Profiles, click Cedar Cove - FG.

- Click Add.

36   Click OK.

The Target Mapping dialog box, Object Name column appears as follows.

37   Click OK twice to close the dialog boxes.

The Cedar Cove corridor with cul-de-sac is reprocessed and is now complete.

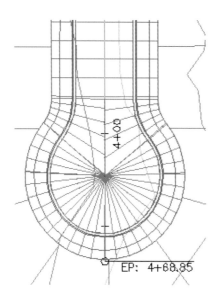

38 Close the drawing. Do not save the changes.

Lesson 29 | Creating Corridor Surfaces

This lesson describes how to create corridor surfaces from corridor models.

Corridor surfaces are useful for design and construction tasks. You can use corridor surfaces to calculate earth cut and fill quantities, label spot elevations and slopes, and generate construction staking data. Corridor surfaces are dynamic and automatically update when the corridor changes.

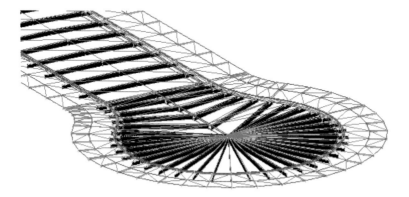

Objectives

After completing this lesson, you will be able to:

- Describe corridor surfaces.

- Describe the process for creating corridor surfaces.

- Create a corridor surface.

About Corridor Surfaces

After you create a corridor model, you can create corridor surfaces. You can use a corridor surface for earth cut and fill volume calculations, finished grade elevation and slope labels, and for calculating pipe network structure rim and invert elevations.

Definition of Corridor Surfaces

A corridor surface represents a finished layer of a corridor design. It is a Civil 3D surface object created from the links or feature lines of a corridor model. Corridor surfaces are reactive surfaces that are displayed in the Surfaces collection on the Prospector tab of the Toolspace window. Corridor surfaces can be displayed, annotated, and sampled just like regular surfaces.

Surface Boundary

You create a corridor surface boundary to contain the triangulation of points within the limits of the boundary. To automatically create a corridor surface boundary, you modify the properties of the corridor.

Examples

For a residential subdivision road corridor, the surface boundary is at the right-of-way line. The following illustration shows a corridor top surface of a residential subdivision road without a boundary defined. The arrows point to the triangulation lines that are outside the surface boundary.

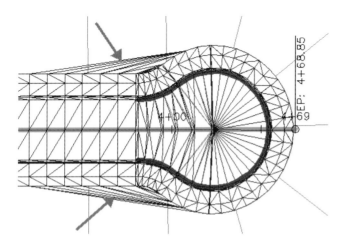

The following illustration shows the same corridor top surface with the boundary defined. In this case, the triangulation lines are contained by the boundary of the corridor surface.

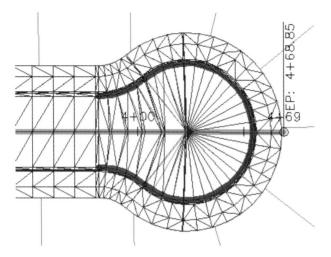

Creating Corridor Surfaces

You use corridor data to create a corridor surface. When you create a corridor surface, you can use either the subassembly links or the corridor feature lines as corridor data. The most common corridor data is the subassembly links.

Creating Corridor Surfaces

To create a corridor surface you modify the corridor properties and use commands available on the Surfaces tab of the Corridor Surface dialog box. The most common corridor surfaces incorporate the Top subassembly links and the Datum subassembly links.

Top Corridor Surface Corridor surfaces using the Top subassembly links can be used for the following:

- Creating finished design contours.

- Labeling finished spot elevations and slopes.

- Creating finished grade construction staking data.

Corridor surfaces using the Datum subassembly links can be used for the following:

- Calculating earth cut and fill quantities.

- Creating subgrade construction staking data.

Guidelines

Keep the following guidelines in mind when you create corridor surfaces.

- You typically create corridor surfaces from either the top links or the datum links.

- You can use the corridor extents to create an outer boundary for the corridor surface.

Exercise 01 | Create a Corridor Surface

In this exercise, you create a top corridor surface for the Cedar Cove corridor.

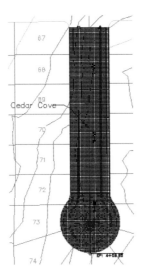

The completed exercise

1   Open *Site Design - Assemblies and Corridors\\I_corridor_surface.dwg*
    (*M_corridor_surface.dwg*).

2   In Prospector, click to expand Corridors. Right- click Cedar Cove. Click Properties.

3   In the Corridor Properties - Cedar Cove dialog box, Surfaces tab:

    • Click Create a Corridor Surface, the first icon on the left.

    • Change the surface Name to **Cedar Cove TOP**.

    • Click in the Surface Style column.

    • In the Pick Corridor Surface Style dialog box, click Grid. Click OK.

    • Under Add Data, verify Data Type is set to Links and Specify Code is set to Top.

    • Click Add Surface Item (plus sign).

    • Click OK.

    You create the surface from the top corridor links. Links connect the points on the
    corridor assembly.

The following illustration shows a surface representing the top of the Cedar Cove corridor. Notice that the surface extends beyond the limits of the corridor near the cul-de-sac. This occurs because a boundary has not been defined for the surface.

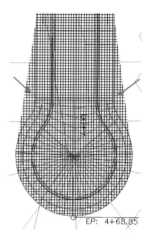

Next, you add boundaries to contain the surface within the limits of the corridor.

4   In Prospector, under Corridors, right-click the Cedar Cove. Click Properties.

5   In the Corridor Properties - Cedar Cove dialog box, Boundaries tab, right-click Cedar Cove TOP. Click Corridor Extents as Outer Boundary. Click OK.

The corridor rebuilds with the Cedar Cove top surface displayed with boundaries.

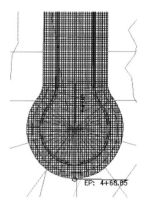

6   Repeat the process and create a corridor surface called Cedar Cove DATUM, using the Datum links.

7   Close the drawing. Do not save the changes.

Chapter 09
Site Design - Grading and Quantities

In this chapter, you design the grading for land parcels adjacent to subdivision roads. You begin by creating grading feature lines from the subdivision road corridor models at the right-of-way locations. The grading feature lines are then offset horizontally and vertically to model the frontage of the building lots and the overall drainage characteristics of the site.

You then create a temporary interim grading surface from the feature lines, which is used to calculate another feature line at the back of parcels location. From this feature line, you grade to the existing ground surface by daylighting with grading objects and slope criteria.

Next, you create a final grading surface that incorporates the daylighting grading objects, the feature lines on the parcels, and the corridor datum surface. The final earthworks volumes are calculated by comparing the final grading surface to the existing ground surface.

Finally, you prepare the engineering and construction plans by adding labels to parcel corners and parcel segments. The elevation labels on the parcel corners reference the final grading surface, as do the grade labels along the parcel segments.

▶ Objectives

After completing this chapter, you will be able to:

- Create feature lines from surfaces and corridors.

- Create interim grading surfaces for grading designs.

- Create final grading surface and calculate volumes.

- Label elevations, slopes, and grades on surfaces.

Lesson 30 | Creating Feature Lines

In this lesson, you work with feature lines to begin the grading process for parcels adjacent to a subdivision road.

Feature lines are similar to 3D polylines. Feature lines have extended capabilities and user interface options that can help you to effectively model proposed grading conditions.

The following illustration shows feature lines created at the corridor right-of-way, at the building pad frontage, and beyond the back of parcels to model the overall flow characteristics of the site.

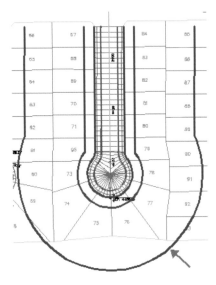

Objectives

After completing this lesson, you will be able to:

- Describe feature lines.

- Create grading feature lines from a corridor.

- Create grading feature lines to begin the grading process for parcels adjacent to a subdivision road.

About Feature Lines

Feature lines are similar to 3D polylines and are used to create proposed grading surfaces. Feature lines have extended capabilities that can be used to simplify grading tasks.

You can edit feature lines with ribbon commands. The Stepped Offset command offsets a feature line horizontally and vertically. There are also commands to join, fillet, and modify feature line elevations and grades. You can also label the elevations of feature line points and the grades of feature line segments. Feature lines can be used as breaklines for proposed surfaces.

Definition of Feature Lines

A feature line is a Civil 3D® object that represents a three-dimensional polylinear element. Feature lines connect a series of geometry and elevation points. You can draw feature lines, create them by converting existing objects, or create them from corridor feature lines. Feature lines are similar to AutoCAD® 3D polylines and store both horizontal and elevation location data. Feature lines can be labeled with grades and elevations.

Example: Grading Feature Line at Right-of-Way

Roads are typically the first part of a subdivision that are designed. Road profiles and elevations account for site topography, site drainage, and optimized cut and fill volumes. Roads therefore form the grading spine for subdivisions. Grading feature lines created from corridor feature lines are the starting point for subdivision grading. These are usually created at the right-of-way locations.

Residential subdivision road corridor models are typically calculated up to the right-of-way locations. After the corridor is calculated, you create a grading feature line from the right-of-way feature line. A grading feature line at the right-of-way location is shown in the following illustration.

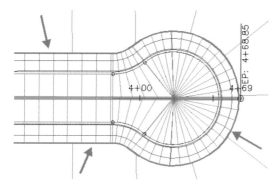

Creating Grading Feature Lines from Corridors

For residential subdivision roads, the corridor model is a three-dimensional representation of the road design up to the right-of-way lines. You grade the remainder of the subdivision by creating grading feature lines from the subdivision road corridors at the right-of-way. Other feature lines are created from corridor feature lines and can be used with grading commands to grade the land parcels adjacent to a corridor model.

Creating Grading Feature Lines Process

A grading feature line created from a corridor is a three-dimensional polylinear element that can be used as a starting point to perform grading tasks beyond the limits of the corridor. Grading feature lines created from a corridor are dynamic and automatically update when the corridor model changes. To edit grading feature lines created from corridors, you can remove the dynamic link.

You create grading feature lines from corridor feature lines using the Feature Line > Create Grading Feature Lines from Corridor command on the ribbon, Home tab, Create Design panel. When a corridor contains multiple baselines, you must individually select the feature lines for each baseline to create grading feature lines.

Guidelines

Keep the following guidelines in mind when you create feature lines.

- Grading feature lines geometry should be simplified when possible. The goal is to minimize the number of elevation and geometry points, without disrupting the integrity of the design.

- Feature lines in the same site interact with each other. If two feature lines overlap, an elevation point is automatically created at the intersection point of the feature lines.

- You can use feature line style to control the appearance of the feature lines.

- You can use the Stepped Offset feature line command to create a new feature line by specifying a horizontal and vertical offset. You can use elevation difference or grade to calculate the elevations of the new feature line.

Exercise 01 | Create and Edit Feature Lines

In this exercise, you create grading feature lines to begin the grading process for parcels adjacent to a subdivision road. You first create a grading feature line from the corridor model at the right-of-way location. The next grading feature line is created at the building frontage location. The final grading feature line models the overall flow characteristics of the site.

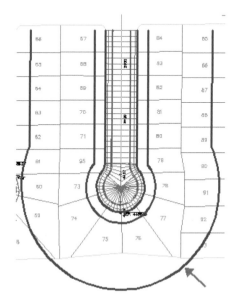

The completed exercise

For this scenario, the builder grades the parcels adjacent to Cedar Cove first.

1 Open *Site Design - Grading and Quantities**I_create_feature_lines.dwg* (*M_create_feature_lines.dwg*).

Next, you create grading feature lines from the corridor object at the right-of-way locations.

2 In Prospector, click to expand Corridors. Right- click Cedar Cove. Click Zoom To.

The Cedar Cove corridor is the corridor on the right.

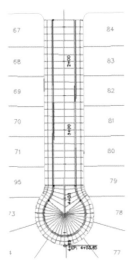

3   On the ribbon, Home tab, Create Design panel, click Feature Line > Create Feature Line
     from Corridor.

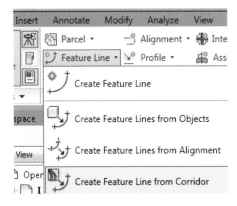

4   At the Select a Corridor Feature Line prompt, for the Cedar Cove corridor, click the feature
     line on the west right-of-way, tangent section.

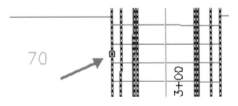

5   In the Create Feature Line from Corridor dialog box:

- For Site, click Grading.

- Select the Name check box.

- For Name, enter **Cedar Cove West ROW**.

- Select the Style check box.

- For Style, click Corridor ROW.

- Click OK.

The west right-of-way feature line is created. The command is still active.

6   At the Select a Corridor Feature Line prompt, click the feature line on the east right-of-way, tangent section.

7   In the Create Feature Line from Corridor dialog box:

- For Site, click Grading.

- Select the Name check box.

- For Name, enter **Cedar Cove East ROW**.

- Select the Style check box.

- For Style, click Corridor ROW.

- Click OK.

The east right-of-way feature line is created. The command is still active.

8   At the Select a Corridor Feature Line prompt, click the feature line on the right-of-way on the cul-de-sac.

9   In the Create Feature Line from Corridor dialog box:

- For Site, click Grading.

- Select the Name check box.

- For Name, enter **Cedar Cove Cul-de-sac ROW**.

- Select the Style check box.

- For Style, click Corridor ROW.

- Click OK.

- Press ENTER.

The cul-de-sac right of way feature line is created. You now have three dynamic feature lines.

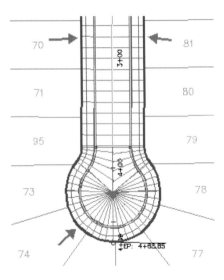

Next, you join the feature lines together. Before doing this you must remove the dynamic links to the corridor.

10   Select the three feature lines. Right-click. Click Remove Dynamic Links.

11   On the ribbon, Modify tab, Design panel, click Feature Line.

12   On the Feature Line tab, click Edit Geometry to display the Edit Geometry panel.

13   In the Edit Geometry panel, click Join.

14   At the Select Object prompt, select the three feature lines consecutively, starting with the west right-of-way and ending with the east right-of-way. Press ENTER.

15   Click the feature line.

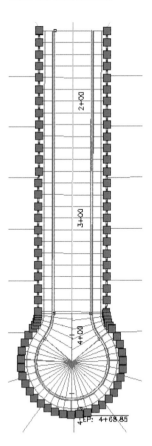

The grips indicate that you now have a single feature line. Next, you rename the feature line in Prospector.

16   In Prospector:

    • Click to expand Sites, Grading.

    • Click Feature Lines.

    • In the item view area, rename the feature line to Cedar Cove ROW.

17   In the drawing area click the feature line. Right- click, click Elevation Editor.

The feature line elevation data is displayed in the Grading Elevation Editor in Panorama.

18   Click in the individual cells. Note the corresponding marker in the drawing area.

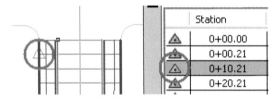

The feature line is no longer dynamic, therefore, in the Grading Elevation Editor, you can change feature line segment grades and elevation points. You can also raise, lower, or flatten the feature line to a common elevation.

19   Click the check mark to close the Grading Elevation Editor.

The feature lines represent the design elevations for the back of sidewalk at the property line locations. Next, you offset the feature lines horizontally and vertically to locate the front of the building pads for the houses on the parcels. You offset the Cedar Cove feature line, horizontally and vertically, using the Stepped Offset command.

20   On the ribbon, Home tab, Create Design panel, click Feature Line > Create Feature Line from Stepped Offset.

21   At the Specify Offset Distance prompt, enter **20 (7 m)**. Press ENTER.

22   At the Select an Object to Offset prompt, click the Cedar Cove ROW feature line.

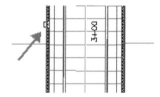

23 At the Specify Side to Offset prompt, click any point to the outside of the selected feature line.

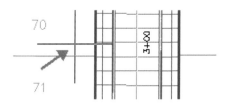

24 At the Specify Elevation Difference prompt, enter **G**. Press ENTER.

25 At the Specify Grade prompt, enter **2**. Press ENTER.

The feature line is offset 20' (7m) at a 2% grade.

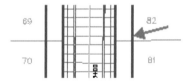

Next, you offset the new feature line to model the overall flow characteristics of the site to promote positive drainage to the back of the parcels.

26 On the ribbon, Home tab, Create Design panel, click Feature Line > Create Feature Line from Stepped Offset.

27 At the Specify the Offset Distance prompt, enter **120' (35 m)**. Press ENTER.

28 At the Select an Object to Offset prompt, select the first offset feature line at the front of the building pad location.

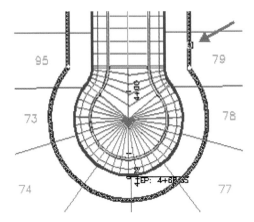

29 At the Specify Side to Offset prompt, click any point to the outside of the selected feature line.

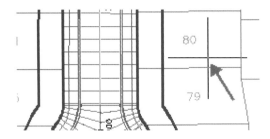

30 At the Specify Grade prompt, enter **D**. Press ENTER.

31 At the Specify Elevation Difference prompt, enter **-1' (-0.3 m)**. Press ENTER twice.

The second offset feature line is created and models the overall flow characteristics of the site. The completed exercise appears as follows.

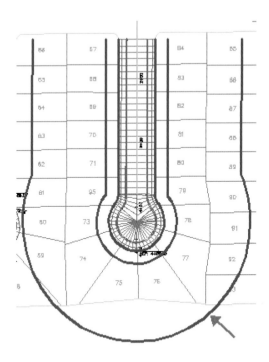

32   Close the drawing. Do not save the changes.

Lesson 31 | Creating Interim Grading Surfaces

This lesson describes the process for creating an interim grading surface that is necessary for solving grading for a subdivision design. You also create a feature line along the back of the Cedar Cove parcels by extracting the elevations from the interim grading surface. This feature line can be used as a grading footprint to daylight to an existing surface.

The creation of an interim grading surface is a procedural step in the creation of a subdivision. It helps you determine design elevations that would otherwise be difficult to calculate.

An interim grading surface that can be used to calculate a feature line along the back of parcels is shown in the following illustration. The feature line along the back of parcels is used for the grading footprint.

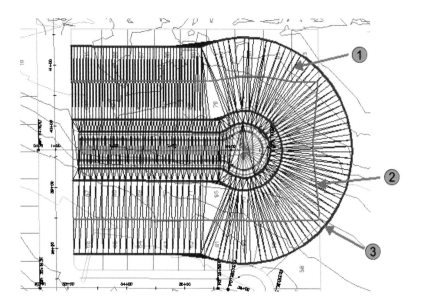

1 Interim grading surface

2 Back of lots

3 Offset feature line

Objectives

After completing this lesson, you will be able to:

- Describe interim grading surfaces.

- Describe the process for creating interim grading surfaces.

- Describe the process for creating feature lines from a surface.

- Create an interim grading surface.

- Create a grading footprint.

About Interim Grading Surfaces

An interim grading surface models the overall flow characteristics of the site, and is used to establish the design elevations at the back of parcels.

Definition of Interim Grading Surfaces

The interim grading surface is an intermediate surface that is created to help you grade elevations at the back of lots.

Facts

To create the interim grading surface, you use the corridor feature lines and the feature lines created with the Stepped Offset command. The surface can be deleted after the elevations at the back of lots are calculated.

Creating Interim Grading Surfaces

The interim grading surface is created from the grading feature lines created from the corridor, and the offset feature lines created with the Stepped Offset feature line command.

Creating Interim Grading Surfaces

Prior to creating the interim grading surface, feature lines exist at the corridor right-of-way and at the setback locations. These feature lines are shown in the following illustration.

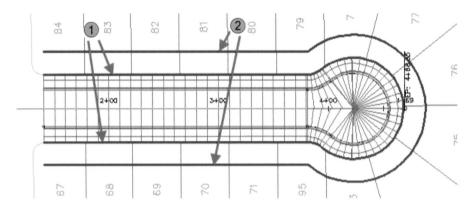

(1) Right-of-Way feature lines

(2) Setback feature lines

The Stepped Offset command is used to offset the setback feature lines to a location beyond the back of the lots. The elevation of the second offset feature lines (beyond the back of lots) is calculated by a negative elevation differential between the setback feature lines and the second offset feature line. This is shown in the following illustration.

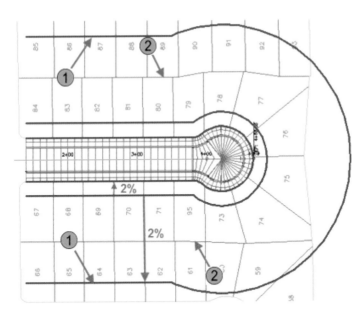

(1) Second Offset feature line

(2) Back of lots

After you create the second offset feature line beyond the back of lots, you then use the feature lines to create the interim grading surface. You create the interim grading surface and then add the feature lines as breakline surface data. This is shown in the following illustration.

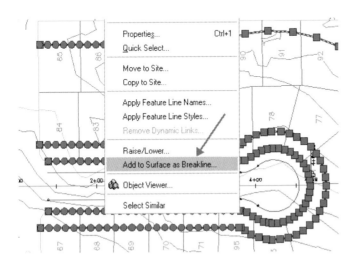

Creating Feature Lines from Surfaces

You create a feature line at the back of parcels that will be used for the grading footprint. The elevations are assigned to the feature line points from the interim grading surface. This process essentially drapes the feature line onto the surface and assigns elevations at the feature line vertices, or elevation points.

Creating Feature Lines from Surfaces

When you assign elevations to a feature line from a surface, you can either assign the elevations at the feature line vertex elevations, or you can add the intermediate grade break points. The intermediate grade break points are calculated at the locations where the feature lines intersect the surface triangulation lines. When you use intermediate grade break points, more elevation vertices are created on the feature line. This is shown in the following illustration.

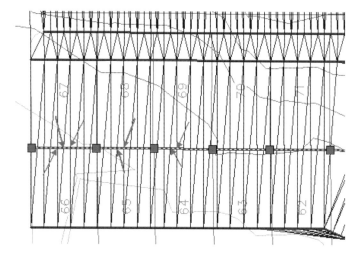

The feature line created at the back of parcels is used as a grading footprint for daylighting to the existing ground surface.

Exercise 01 | Create an Interim Grading Surface

In this exercise, you create an interim grading surface. The creation of an interim grading surface is an intermediate step that enables you to calculate the elevations at the back of the parcels adjacent Cedar Cove.

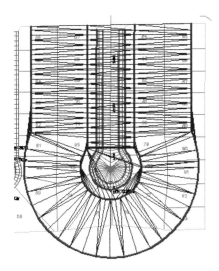

The completed exercise

You begin by extending the offset feature lines to property lines to the north. To do this, you use the AutoCAD Extend command.

1   Open the file \Site Design - Grading and Quantities\I_interim_grading_surface.dwg (*M_interim_grading_surface.dwg*).

    First, you extend the feature lines to the property lines using the AutoCAD extend command.

2   On the command line, enter **EX**. Press ENTER.

3   When prompted to select boundary edges, select the north parcel segments for Parcels 66 and 67, and 84 and 85. Press ENTER.

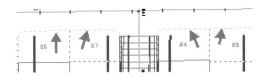

4   When prompted to select the object to extend, click near the ends of the two offset feature lines on either side of Cedar Cove. You click four times. Press ENTER.

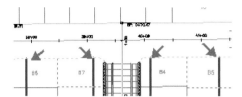

The offset feature lines are extended north to the property lines. The grade of the last feature line segment is maintained. Feature line creation required for the interim grading surface is complete.

Next, you create the interim grading surface.

5   In Prospector, right-click Surfaces. Click Create Surface.

6   In the Create Surface dialog box:

   • For Name, enter **Interim Grading**.

   • For Description, enter **Interim Grading Surface**.

   • For Style, click _Triangles.

   • Click OK twice to close the dialog boxes.

Next, you add the feature lines to the interim grading surface as breakline data.

7   In the drawing area:

   • Select any feature line.

   • Right-click, click Select Similar.

   • Right-click. Click Add to Surface as Breakline.

8   In the Select Surface dialog box, click Interim Grading. Click OK.

9   In the Add Breaklines dialog box, click OK.

The interim grading surface is created.

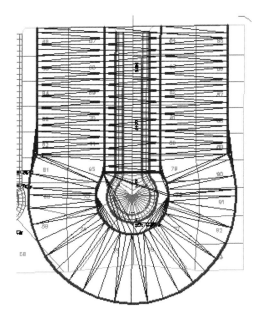

10 Close the drawing. Do not save the changes.

Exercise 02 | Create a Grading Footprint

In this exercise, you create a feature line along the back of the Cedar Cove parcels that will be used as a grading footprint. The elevations of the feature line points are calculated from the interim grading surface. You also combine the corridor datum surface data to begin the creation of the final grading surface.

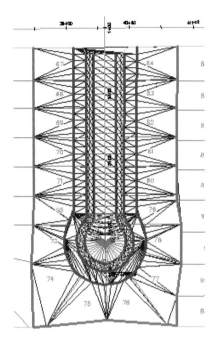

The completed exercise

You begin by suppressing the display of the interim grading surface.

1   Open *Site Design - Grading and Quantities\I_grading_footprint.dwg* (*M_grading_footprint.dwg*).

2   In the drawing area:

   • Click the interim grading surface.

   • Right-click. Click Surface Properties.

3   In the Surface Properties dialog box, Information tab, for Surface Style, click _No Display. Click OK.

Next, you create a feature line that runs along the back of parcels adjacent to Cedar Cove. You begin by setting the running object snaps to endpoint.

4   Turn OSNAPS on and set the default OSNAP setting to Endpoint.

5   On the ribbon, Home tab, Create Design panel, click Feature Line > Create Feature Line.

6   In the Create Feature Lines dialog box:

- For Site, click Grading.

- Select the Name check box.

- For Name, enter **Cedar Cove Back of Lots**.

- Select the Style check box.

- For Style, click Grading.

- Click OK.

7   When prompted to specify start point, click and snap to the north endpoint of the parcel segment between parcels 66 and 67.

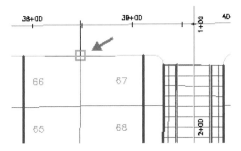

8   When prompted to specify elevation, press ENTER.

9   When prompted to specify the next point, click and snap to the back of parcel endpoint of the parcel segment between Parcels 67 and 68.

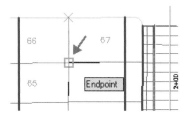

10 When prompted to specify elevation, press ENTER.

11 When prompted to specify the next point, click and snap to the back of parcel endpoint of the parcel segment between Parcels 68 and 69.

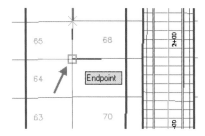

12 When prompted to specify elevation, press ENTER.

13 Repeat steps 11 and 12. Select the endpoint of the back of all parcel segments and deflection points surrounding Cedar Cove. Press ENTER when finished.

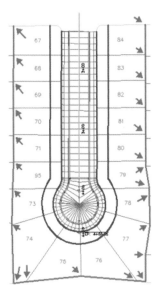

The feature line elevations are all 0.

Next, you assign elevations to the feature line from the interim grading surface. This is equivalent to projecting the feature line to the interim grading surface.

14 On the ribbon, Modify tab, Edit Elevations panel, click Elevations from Surface.

15 In the Set Elevations from Surface dialog box:

- Select Interim Grading.

- Clear the Insert Intermediate Grade Break Points check box.

- Click OK.

16 When prompted to select object, select the feature line on the back of the Cedar Cove lots. Press ENTER.

17 Select the same feature line. Right-click. Click Elevation Editor.

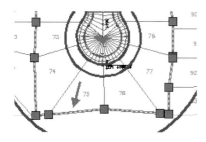

The elevations and grades are displayed in the Grading Elevation Editor in Panorama. These elevations and grades were calculated from the interim grading surface.

18 Close Panorama.

Next, you add the new feature line to the interim grading surface. First, you display the interim grading surface.

19 In Prospector, Expand Surfaces. Right-click Interim Grading. Click Surface Properties.

20 In the Surface Properties - Interim Grading dialog box, Information tab, for Surface Style, click
    _Triangles. Click OK.

    The interim grading surface is displayed as triangles.

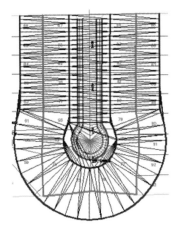

21 In the drawing area:

    • Select the feature line on the back of the Cedar Cove parcels.

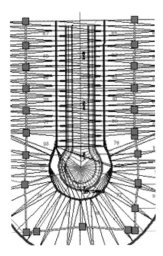

    • Right-click. Click Add to Surface as Breakline.

22 In the Select Surface dialog box, click Interim Grading. Click OK.

23 In the Add Breaklines dialog box, click OK.

The feature line data is added to the surface and the surface triangulation updates.

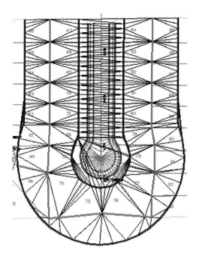

Next, you add the corridor datum data to the interim grading surface. Earthwork volumes for the site design include the grading required for the roads. The corridor datum is represented by the Cedar Cove Datum surface. This surface is provided for you. You use the Paste Surface command to add the corridor datum surfaces.

24 In Prospector, expand Surfaces, Interim Grading, Definition. Right-click Edits. Click Paste Surface.

25 In the Select Surface to Paste dialog box, click Cedar Cove Datum. Click OK.

The surface now includes the datum data for the Cedar Cove corridor.

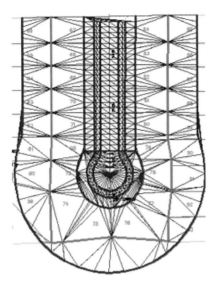

Next, you erase the outermost offset feature line. This results in a required update to the interim grading surface.

26 In the drawing area, select the surface. Right- click. Click Display Order > Send to Back.

27 Click the outermost offset feature line.

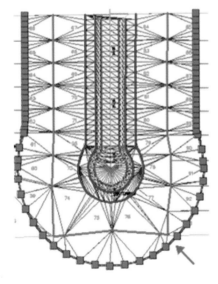

Chapter 09 | Site Design - Grading and Quantities

28 Press DELETE.

29 In Prospector, note that the interim grading surface is displayed with an exclamation mark.

The exclamation mark indicates that the surface requires updating. The outermost offset feature line you deleted was a breakline for the Interim Grading surface. Deleting surface data therefore warrants a rebuild of the surface.

30 In Prospector, right-click Interim Grading. Click Rebuild.

The Interim Grading surface has been rebuilt and now excludes the outermost feature line. The completed exercise appears as follows.

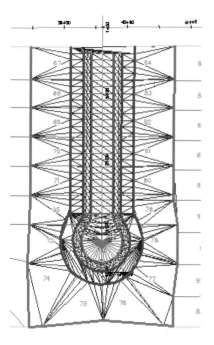

31 Close the drawing. Do not save the changes.

Lesson 32 | Creating Final Grading Surfaces and Calculating Volumes

This lesson describes how to create grading objects and calculate volumes between two surfaces. You use the grading objects to create a surface to generate volume calculations that estimate the required materials for your design. As with any drawing object, the grading design and any related objects are automatically updated whenever you make a change to the grading.

A final grading surface is shown in the following illustration.

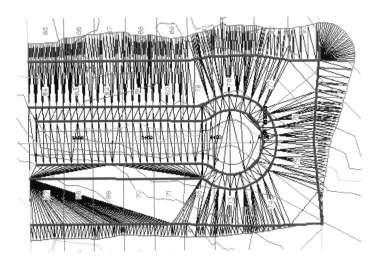

Objectives

After completing this lesson, you will be able to:

- Describe grading objects.

- Describe the process for creating grading objects.

- Explain the characteristics and function of TIN volume surfaces.

- Explain how you calculate earthwork volumes.

- Create a grading object to model the daylighting from the back of parcels feature lines to the existing ground surface.

- Create final grading surface and calculate the total earthworks volumes.

About Grading Objects

Grading objects and grading groups are integral to grading a surface. Grading objects are created in the engineering phase of a land development project and used in the subsequent development phases. Criteria and styles can be assigned to a grading object or group, and surfaces are created from grading objects or groups.

A grading object is shown in the following illustration.

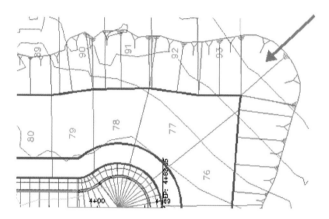

Definitions

Grading Objects

A grading object represents an existing or proposed design feature. It represents a projected slope to either a surface, elevation, or an offset. A grading object typically consists of:

- The baseline, which can be a feature line or a lot line.

- The target, which can be a surface, a distance, an elevation, or a relative elevation.

- Projection lines that define the direction of the grading.

- The face, which is the area enclosed by the baseline, the target line, and the projection lines.

After it is created, you apply grading criteria and styles to the grading object. A grading object must be assigned to a grading group.

Grading Groups

Grading groups are part of Civil 3D sites. Grading groups organize grading objects, contain volume data for the grading objects, and contain the configuration that combines grading objects into a surface. You create a surface from multiple grading objects within a grading group.

Example

If you create grading objects that represent a runoff and a drainage ditch, you can create the two gradings as part of the same grading group. If the group is configured to create a surface, the two objects combine to create the surface in the drawing.

Creating Grading Objects

To create grading objects, you do the following:

- Select/create a grading group.

- Select a grading footprint feature line.

- Select a grading criteria.

- Select a target surface (if criteria targets a surface).

- Specify grading parameters.

Creating Grading Objects

To create a grading object you create a grading group, select a grading criteria, and, if the grading criteria involves a surface, you specify the target surface. You create a grading object from a feature line, which is sometimes referred to as the grading footprint. The grading criteria indicates how the grading is calculated.

You create grading with the Create Grading command on the Creation Tools toolbar, which is shown in the following illustration.

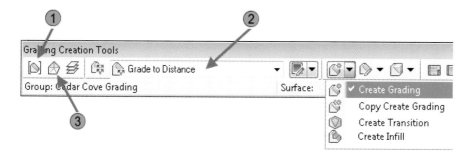

1. Set Grading Group

2. Grading Criteria

3. Target Surface

When you create the grading group, you have the option to automatically create a grading surface and a comparison surface for volumes as part of the grading group. If you choose to bypass this option, you can perform these steps later.

After you create the grading group you specify the target surface, which is the Existing Ground surface. You finally select the grading criteria, which is Grade to Surface and create the grading and specify the slopes. This is shown in the following illustration.

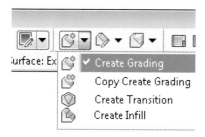

About TIN Volume Surfaces

Earthwork volume quantities are vital components in the design of a subdivision. Earthwork operations are one of the most expensive elements of a design. Therefore it is desirable to "balance" the earthwork. Balance is achieved when the material to be cut equals the material to be placed as fill. Civil 3D combines two surfaces to create a TIN volume surface, which includes as part of its properties the cut and fill quantities.

The Create Surface dialog box, with the TIN Volume Surface type selected is shown in the following illustration. Note the base and comparison surfaces, and the cut and fill factors.

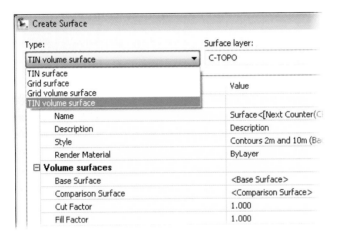

Definition of TIN Volume Surfaces

A triangulated irregular network, or TIN, is a surface model consisting of data points (vertices) connected by 3D lines (TIN lines) to form three-dimensional irregularly shaped triangular faces. These triangular faces are collectively called a TIN. TINs are used to model existing ground surfaces, proposed surfaces, subsurfaces (like bedrock), and water surfaces. A volume TIN surface is a TIN resulting from the comparison of two different TINs, such as an existing ground and a proposed ground surface.

Examples of TIN Volume Surfaces

TIN volume surfaces are used to compare and calculate:

- Proposed to existing surfaces.

- Detention basin storage volumes.

- Detention basin earthwork cut volumes.

- Berm and stockpile volumes.

Calculating Earthwork Volumes

You can calculate earthwork volumes in several ways, two of which are introduced here. The
first method employs a simple calculation and displays the results in the Panorama window. The
second method is more complex and uses the EG surface and the prefinal top surface to create a
composite surface for calculating volume. The advantage to the second method is that the volume
surface is similar to a standard TIN surface and is an object in your drawing. Therefore, its styles
can be modified, analyzed, and used for exhibits, such as cut and fill maps. Both methods give you
the same cut, fill, and net volume results.

Process: Calculating Earthwork Volumes (Method 1)

The following steps outline the process for calculating earthwork volumes using the simple
calculation method.

1   Create a new volume:

- Use the Surfaces menu or Panorama window.

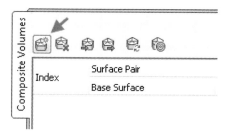

2   Select the base and comparison surfaces.

   • Select the base surface (EG).

   • Select the top surface (Prefinal Top).

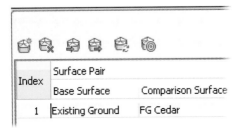

3   Calculate the volumes (done automatically by AutoCAD Civil 3D):

   • Examine the calculated cut, fill, and net volumes.

Surface Pair		Volume	
Base ...	Comp...	Cut	Fill
Existin...	FG Ce...	12773.90 Cu. Yd.	3767.58 Cu. Yd.

Process: Calculating Earthwork Volumes (Method 2)

The following steps outline the process for calculating earthwork volumes using the composite surface method.

1   Create a new surface and set base and top surfaces.

   • Create the TIN volume surface.

   • Select the base surface.

   • Select the comparison surface.

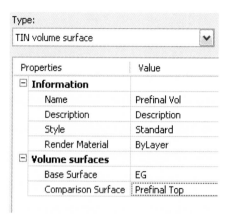

Type:

TIN volume surface

Properties	Value
Information	
Name	Prefinal Vol
Description	Description
Style	Standard
Render Material	ByLayer
Volume surfaces	
Base Surface	EG
Comparison Surface	Prefinal Top

2   Calculate volumes.

• Examine calculated cut, fill, and net volumes.

Statistics	Value
General	
TIN	
Volume	
Base Surface	EG
Comparison Surface	Prefinal Top
Cut volume (unadjusted)	24640.79 Cu. Yd.
Fill volume (unadjusted)	33803.99 Cu. Yd.
Net volume (unadjusted)	9163.20 Cu. Yd. <Fill>

Exercise 01 | Create Grading Objects

In this exercise, you create a grading object to model the daylighting from the back of parcels feature lines to the existing ground surface.

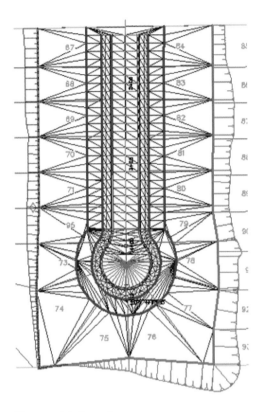

The completed exercise

1   Open *Site Design - Grading and Quantities**I_grading_object.dwg* (*M_grading_object.dwg*).

2   On the ribbon, Home tab, Create Design panel, click Grading > Grading Creation Tools.

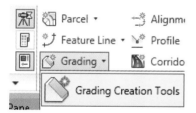

The Grading Creation Tools toolbar is displayed. Next, you select a site then create a grading group.

3   On the Grading Creation Tools toolbar, click Set the Grading Group.

4   In the Site dialog box, select Grading. Click OK.

5   In the Create Grading Group dialog box, for Name, enter Cedar Cove Grading. Click OK.

     Next, you select the surface that you grade to. This is the existing ground surface.

6   On the Grading Creation Tools toolbar, click Set the Target Surface.

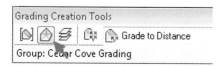

7   In the Select Surface dialog box, click Existing Ground. Click OK.

     You are now ready to begin the grading calculation.

8   On the Grading Creation Tools toolbar, for Grading Criteria, click Grade to Surface.

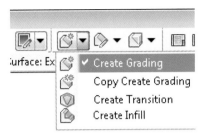

9   Click Create Grading.

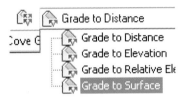

10   When prompted to select the feature, click the feature line on the back of the parcels adjacent to Cedar Cove.

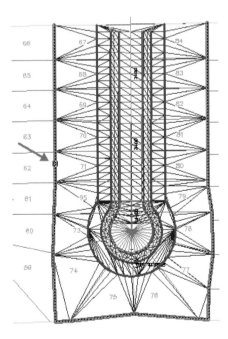

11 When prompted to select the grading side, click any location outside of the feature line.

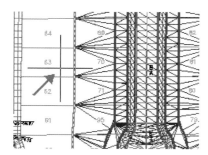

12 When prompted to apply to entire length, press ENTER.

13 When prompted for cut format, press ENTER.

14 When prompted for cut slope, enter **5**. Press ENTER.

15 When prompted for fill format, press ENTER.

16 When prompted for sill slope, enter **5**. Press ENTER.

17 Press ENTER to end the command.

The grading object is calculated and displayed in the drawing area.

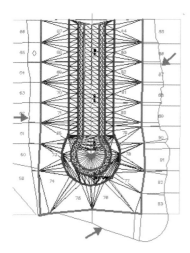

The colors of the grading are green on the left side (fill slopes) and red on the right side (cut slopes). Fill and cut grading styles are used to portray cut slopes and fill slopes differently.

18 In the drawing area:

- Click any grading slope projection line.

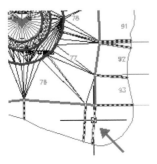

- Right-click, click Grading Properties.

19 In the Grading Properties dialog box, notice the assignment of the Cut Slope Display and Fill Slope Display grading styles.

Next, you modify the grading styles.

20 Click the second down arrow to the right of Cut Slope Display. Click Edit Current Selection.

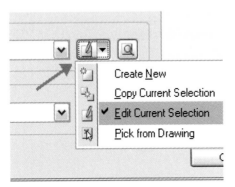

21 In the Grading Style - Cut Slope Display dialog box, Slope Patterns tab, select the Slope Pattern checkbox. Click OK.

22 Repeat steps 20 and 21. Modify the Fill Slope Display grading style to turn on slope patterns.

23   Click OK to close Grading Properties dialog box.

The display of the grading object is updated to include slope patterns for cut and fill.

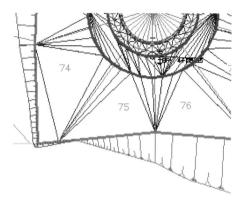

Note: You can also modify grading styles from the Settings tab of the Toolspace window.

Next, you modify the slopes.

24   Close the Grading Creation Tools toolbar.

25   In the drawing area, select any grading projection line. Right click, click Grading Editor.

The Panorama window displays the Grading Editor.

26   In the Grading Editor:

- For Cut Slope Projection (up), Slope, change the value to **3**.

- For Fill Slope Projection (down), Slope, change the value to **3**.

The grading object recalculates and projects a 3:1 slope to the surface.

27 The completed exercise drawing appears as follows.

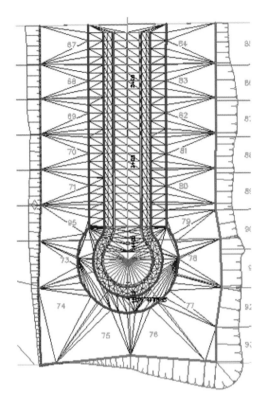

Close the drawing. Do not save the changes.

Exercise 02 | Create Final Grading Surface and Calculate Volumes

In this exercise, you calculate the total earthworks volumes.

Surface Pair		Volume	
Base ...	Comp...	Cut	Fill
Existin...	FG Ce...	12773.90 Cu. Yd.	3767.58 Cu. Yd.

The completed exercise

1   Open \Site Design - Grading and Quantities\I_surface_and_volumes.dwg
    (M_surface_and_volumes.dwg).

2   In Prospector, Expand: Sites, Grading, Grading Groups. Right-click Cedar Cove Grading.
    Click Properties.

3   In the Grading Group Properties - Cedar Cove Grading dialog box, Information tab, select
    Automatic Surface Creation check box.

4   In the Create Surface dialog box:

    • For Name, enter **FG Cedar**.

    • For Description, enter **Finished Ground Cedar Cove**.

    • For Style, click _Grid.

    • Click OK.

5   Click OK to close Grading Group Properties.

A surface is created from the grading object. Notice the new surface in the drawing area.

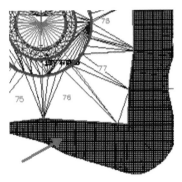

Next, you add the interim grading surface data to the FG Cedar surface. You begin by suppressing the display of the interim grading surface.

6   In Prospector, click to expand Surfaces. Right- click Interim Grading. Click Surface Properties.

7   In the Surface Properties - Interim Grading dialog box, Information tab, for Surface Style, click _No Display. Click OK.

8   In Prospector, under Surfaces, FG Cedar, and Definition. Right-click Edits. Click Paste Surface.

9   In the Select Surface to Paste dialog box, click Interim Grading. Click OK.

The final FG Cedar surface is created.

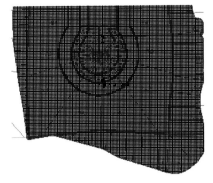

Next, you calculate the volumes.

10   On the ribbon, Analyze tab, Volumes and Materials panel, click Volumes > Volumes.

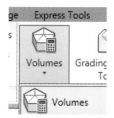

The Composite Volumes calculator is displayed in Panorama. Next, you specify the two surfaces used for the volume calculation.

11   In Panorama:

- Click Create New Volume Entry.

- For Base Surface, click Existing Ground.

- For Comparison Surface, click FG Cedar.

The volumes are calculated and displayed.

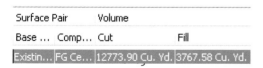

Surface Pair		Volume	
Base ...	Comp...	Cut	Fill
Existin...	FG Ce...	12773.90 Cu. Yd.	3767.58 Cu. Yd.

12   Close Panorama.

13   Close the drawing. Do not save the changes.

Lesson 33 | Labeling Final Grading Surface

This lesson describes how to label elevations, slopes, and grades on surfaces.

Surface labels can help you with the design process because they automatically update when the surface changes. Additionally, it is a common practice to label finished grade surfaces to convey design information to contractors and other interested parties.

The following illustration shows surface labels on a finished grade surface.

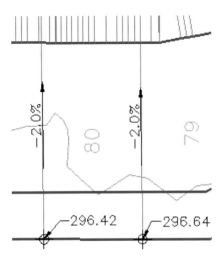

Objectives

After completing this lesson, you will be able to:

- Explain the benefits and drawbacks of two commonly used methods for surface annotation.

- Describe how to create spot elevation labels.

- Create spot elevation labels and grade labels for the design grading surface.

Surface Annotation

In order to convey the engineering grading design intent, surfaces that are to be constructed in the field must be properly labeled or annotated. Elevations at key locations on the surface must be clearly identified. Additionally, the process of annotation itself must be both an accurate and relatively simple process. Civil 3D provides several methods for annotating a surface.

The following illustration shows elevations for the high point (HP), building setback line (BSL), and top of curb (TC).

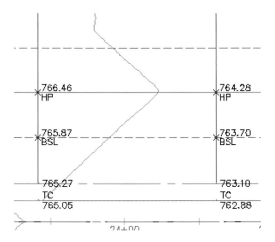

Surface Annotation

Surface annotation can be accomplished in many ways. A common method is to create dynamic surface labels that automatically update when the underlying surface model changes. A second method is to create a point object that can receive its elevation from an underlying surface. While points can be used as part of a surface definition, they do not automatically update if the surface changes from which their elevation was derived.

Examples

Surface annotation is useful for labeling:

- Top of curb elevations.

- High/low point elevations.

- Lot corner elevations.

- Swale centerlines.

- Detention basin outfall elevations.

Creating Spot Elevation Surface Labels

You refine the grading of the individual lots by setting spot elevations at points of interest on the lot lines. You use surface labels to annotate spot elevations at various locations. You create points at the lot line intersections with the building setback line, and at high points on the side yard lot lines. The labels assist in relaying design intent to the review agencies and contractors, while the points add detail to the grading design, moving the process toward a more complete final design.

Process: Creating Spot Elevation Surface Labels

The following steps outline the process for creating spot elevation surface labels.

1   Add labels:

- Set default point descriptions.

- Set running OSNAPS to endpoint and intersection.

- Run the Create Random Points (on Surface) command.

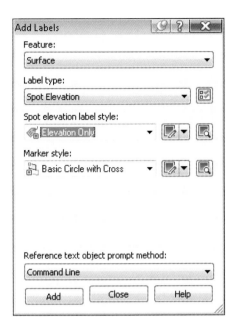

2   Set points at the building setback lines and high points.

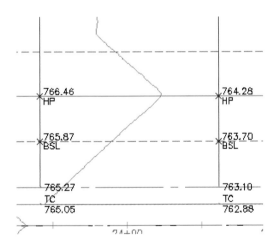

Guidelines

Keep the following guidelines in mind for creating spot elevation and grade labels.

- You can use the Apparent Intersection running object snap to select the location to be labeled. However, doing this may require two clicks for each lot line. On a project of this size, this action might not be overly time-consuming, but on larger projects it is.

- You can set points on a surface where the elevation of the points automatically sets to the underlying surface elevation. However, when the underlying surface is modified, the point elevations remain static. Therefore, you should use spot elevation surface labels because they are updated when the surface changes, making it that much easier to keep the annotation and the design elements synchronized. In anticipation of future revisions to the road profile and, thus, the lot grading, use a combination of both points and spot elevation surface labels.

- You can label grades and elevations on feature lines using line labels. This is helpful when you are grading a site using feature line and need to visualize elevation and grade data for the design.

Exercise 01 | Label Final Grading Surface

In this exercise, you create spot elevation labels and grade labels for the design grading surface.

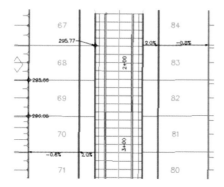

The completed exercise

1   Open \Site Design - Grading and Quantities\I_surface_labels.dwg (M_surface_labels.dwg).

You begin by suppressing the display of the FG Cedar surface.

2   In Prospector, click to expand Surfaces. Right- click FG Cedar. Click Surface Properties.

3   In the Surface Properties - FG Cedar dialog box, Information tab, for Surface Style, click _No Display. Click OK.

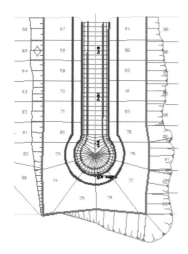

Next, you label spot elevations on the lot corners.

4   On the ribbon, Annotate tab, Labels & Tables panel, click Add Labels > Surface > Add Surface Labels.

5   In the Add Labels dialog box:

- For Label Type, click Spot Elevation.

- For Spot Elevation Label Style, click Elevation Only.

- Click Add.

6   When prompted to select a surface, press ENTER.

7   In the Select a Surface dialog box, click FG Cedar. Click OK.

8   When prompted to select a point, snap to the west endpoint of the parcel segments between parcels:

- 67 and 68

- 68 and 69

- 69 and 70.

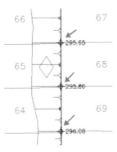

9   Press ENTER when finished.

10 Select the label at the end of the parcel segment between Parcels 67 and 68.

A square-shaped and a diamond-shaped grip are displayed. The square grip is the label location grip and controls the position of the label. The diamond grip is the label anchor grip and is the location being labeled.

11   In the drawing:

- Click the diamond-shaped label anchor grip.

- Move the mouse around the drawing area.
  Notice the preview of the elevation label changes with the location.

- Click the east endpoint of the parcel segment between Parcels 67 and 68 to reposition the label anchor.

- Press ESC.

The label updates to reflect the elevation of the selected location. You can also copy and move spot elevation labels using AutoCAD Copy and Move commands.

IMAGE15

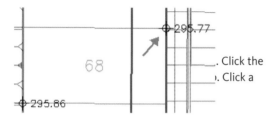

location to the left to reposition the label.

The label is repositioned relative to the label anchor and a leader is inserted. The label adopts its dragged state display property. Note that you can add additional leader vertices using the Add leader vertex grip.

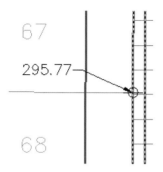

Next, you add slope labels. The Add Labels dialog box should still be open.

13 In the Add Labels dialog box:

- For Label Type, click Slope.

- For Slope Label Style, click Percent.

- Click Add.

14 When prompted to select a surface, press ENTER.

15 In the Select a Surface dialog box, click FG Cedar. Click OK.

16 When prompted to create slope labels, enter **T**. Press ENTER.

17 When prompted to select first point, using the apparent intersection OSNAP, snap to the apparent intersection of the parcel segment between parcels 70 and 71, and the setback feature line.

Tip: When using the apparent intersection OSNAP, click the feature line first and then the parcel segment. You may need to zoom closer.

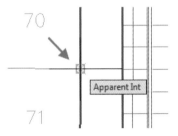

18 When prompted to select second point, snap to the west endpoint of the parcel segment between parcels 70 and 71.

The grade between the two points is labeled.

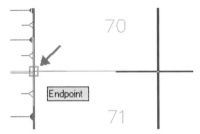

Next, you label the grade on the parcel segment between the setback and the front of the parcel. The label command is still active.

19   When prompted to select first point, using the endpoint OSNAP, snap to the east endpoint of the parcel segment between Parcels 70 and 71.

20   When prompted to select second point, using the apparent intersection OSNAP, snap to the apparent intersection of the setback feature line and the parcel segment between parcels 70 and 71.

A second dynamic slope label is created.

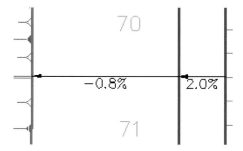

21   Repeat the procedure to label the parcel segments grades on the FG Cedar surface between Parcels 84 and 83.

22   The completed exercise drawing appears as follows.

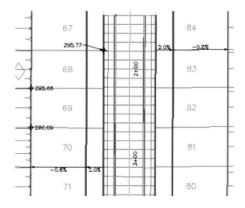

23   Close the drawing. Do not save the changes.

Chapter 10
Site Design - Pipes

In this chapter, you layout and design a storm sewer pipe network for a new subdivision design. You can layout storm sewers, sanitary sewers, watermain systems, electrical ducting, and any other types of subsurface utilities using pipe network functionality in Civil 3D®. The lessons in this chapter guide you through the process for creating, editing, and annotating a storm sewer pipe network. You also use the Hydraflow Storm Sewers Extension to size pipes and calculate invert elevations from flow data.

Pipe networks are created from pipe and structure parts that exist in a parts list. You create a parts lists containing the pipe and structure parts that you would regularly use for your company or client projects. Pipe and structure parts are added to the parts list from the pipes and structures parts catalog. You typically create a separate parts list for storm sewer, sanitary sewer, and watermain systems. You save the parts list in your company or client drawing template (DWT) file to provide users access to the standardized pipe and structure parts that you regularly use on your projects.

When you create a pipe network, predefined rules are applied to calculate the invert elevations and the slopes of the pipes in pipe network. You create a pipe network by selecting the locations of the structure and pipes in plan view. Structure rim elevations are calculated from a finished design surface. Invert elevations are automatically calculated using a combination of minimum/maximum cover and minimum/maximum slope criteria.

Pipe network invert elevations and slopes can be edited either graphically or in a table. When you make changes to elevations and slopes, the graphical display and the annotation of the pipe networks automatically update. You can also use grips to edit the position of pipes and structure graphically in the drawing area. When you move a structure in plan view, the display of the pipe network in the profile view automatically updates.

You use the Hydraflow Storm Sewers Extension to calculate pipe diameters and invert elevations based on inputting flow data. You export a pipe network layout from Civil 3D. Then, you import the data to Hydraflow Storm Sewers Extension, input the flows, and calculate pipe sizes and invert elevations. Finally, the updated data is exported back to Civil 3D.

Objectives

After completing this chapter, you will be able to:

- Create a pipe network that represents a storm sewer design in a subdivision.

- Draw pipe network parts in profile views and edit the pipe network.

- Label pipe networks in plan and profile view.

- Use the Hydraflow Storm Sewers Extension to calculate pipe diameters and invert elevations.

Lesson 34 | Creating Pipe Networks

This lesson describes pipe networks and how you add pipes and structures to a pipe network in plan view.

You create a pipe network to model storm sewer, sanitary sewer, and watermain systems. By creating a 3D model of a pipe network, you can quickly explore different design alternatives and check for interferences with other subsurface features.

The following illustration shows a pipe network in plan, profile, and 3D views. The arrows indicate the pipe network.

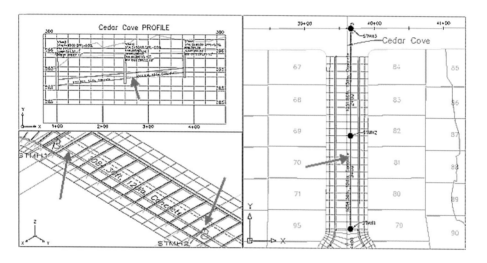

Objectives

After completing this lesson, you will be able to:

- Describe the characteristics and function of pipe network objects.

- Describe the tools that you use to create a pipe network.

- Describe the three pipe network part creation modes.

- List the steps for creating pipe networks.

- Create a storm sewer pipe network.

About Pipe Networks

A pipe network is a connected system of pipes and structures, which can be governed by rules. Its properties define the relationship between the network parts and other objects, such as surfaces and alignments.

The following illustration shows the structures for a storm sewer pipe network in Prospector.

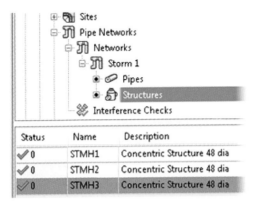

Definition of Pipe Networks

A pipe network is a system of related pipe and structure parts with properties that define relationships among the network parts, alignments, and surfaces.

Pipe Network Relationships

A pipe network also defines the relationship between the network parts and the following objects and resources.

Option	Description
Surface	If you configure your pipe network to reference a surface, the elevation data of the surface is used to determine the vertical sizing and placement of network parts. Sizing and placement of parts are calculated using the surface data and design rules for individual parts.
Alignment	A pipe network can take its station data from an associated alignment. Label your network parts to take the station value from the associated alignment.
Labels	You can configure your pipe network to automatically add labels of the selected type to all pipes and structures that you add to the network. You can also add labels later.

You configure a pipe network when you create it. You can modify its configuration later using the Pipe Network Properties dialog box or the Network Layout Tools toolbar.

Example of a Simple Pipe Network

The following illustrations show the development of a simple pipe network.

In most cases, a pipe network design starts with another drawing object. In this example, the starting point is an alignment, as shown in the following illustration. The elevations of the pipe network parts created are determined from a surface. The surface can be an existing surface or a corridor surface representing the finished grade.

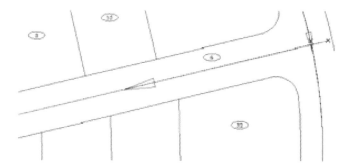

When you use another drawing object as a reference, you can use drawing tools such as Object Snap, transparent commands, and tooltips to help you select locations for your pipe network parts. In the following illustration, two structures are selected, creating a network segment with two structures connected by a pipe.

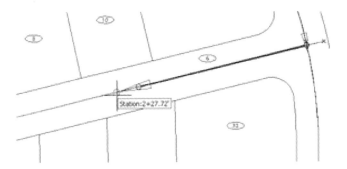

The parts you create are added to the pipe network object in the Prospector tab tree view. When you select the Pipes or Structures items, their properties are displayed in the item view. In the following illustration, the three structures added to the network are displayed.

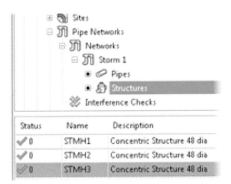

Pipe Network Creation Tools

There are several tools you use to create pipe networks:

- Pipe Network Catalog

- Parts List

- Part Rules

Network Layout Tools

Pipe Network Catalog

The pipe network catalog installs with the software and is external to drawings. It contains all of the available structure and pipe types.

Parts List

You create a parts list and include only those parts that you would regularly use to create the pipe network. Your parts list is created based on the parts contained in the pipe network catalog, and is saved in the drawing template DWT file. Parts lists are useful for organizing pipe network parts. You create a separate parts list for storm sewers, sanitary sewers, and watermains.

Pipe and Structure Rules

Pipe and structure rules govern how the engineering details of a pipe network are calculated when the pipe network is initially created, or when you choose to apply them. The rules also affect how the pipe network parts behave when they are moved or edited. Pipe and structure rules can be associated with the individual pipe and structure parts you add to the Parts List. Pipe rules and structure rules are created independent of each other and are organized into rule sets.

Pipe rules account for the following:

- Cover and slope: Minimum/maximum cover and minimum/maximum slopes

- Cover only: Minimum and maximum cover

- Length check: Minimum and maximum pipe lengths

- Pipe-to-pipe match: Pipe drop and connection location (invert, obvert, center) between adjoining pipes

Structure rules account for the following:

- Maximum pipe size check: Maximum pipe diameter a structure can accommodate

- Pipe drop across structure: Change in elevation between invert in and invert out

- Sump depth: Specify structure sump depth

Pipe Network Part Creation Modes

You can draft all of the components for a utility network in a single operation using pipe network part creation modes. Depending on the requirements of your pipe network project and your design method, you select one of the following pipe network part creation modes:

- Pipes and Structures

- Pipes Only

- Structures Only

Pipe Network Part Creation Modes

The pipe network part creation mode determines which network parts are added as you create a pipe network.

Option	Description
Pipes and Structures	Use this mode to create network parts by selecting locations for a series of structures. Pipes that connect the structures are created automatically. This mode is useful for quickly creating a simple network such as a "cross-country" branch of a sewer or sanitary system.
Pipes Only	Use this mode to create a network of pipes that are not connected by structures. If you have already created structures, you can use the Pipes Only mode to create connecting pipes to complete or add to the network.
Structures Only	Use this mode to create only structures in your network. You can add pipes to the structures later. For example, you can place all the catch basins required by your project first and add pipes later to create the configuration that is most efficient and uses the smallest quantity of materials.

Null Structures

When you create pipes that connect without structures, a null structure is created. A null structure has no function except to connect two pipes. Null structures appear as simple objects in the drawing area and are listed in the Prospector tab tree view.

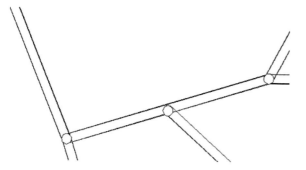

Pipes connected by null structures (2D)

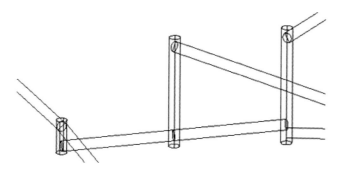

Pipes connected by null structures (3D)

Examples of Using Pipe Network Part Creation Modes

The following illustrations show examples of the Pipes Only and Structures Only pipe network part creation modes.

In your design project, if you need to show one or more pipes in another network that may conflict or interfere with your project, you can use the Pipes Only mode to create the required pipes without having to create connected structures. You can also use Pipes Only mode when you need to create a single, unattached pipe, such as a pipe used as a culvert under a road crossing, as shown in the following illustration.

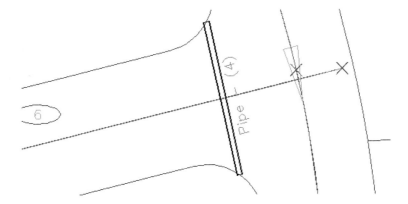

In the following illustration, the designer drew all the required structures without pipes using Structures Only mode.

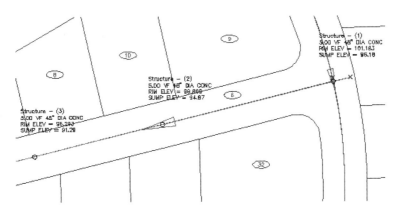

You can add pipes to the design later using Pipes Only mode to make efficient use of space and materials. The completed design is shown in the following illustration.

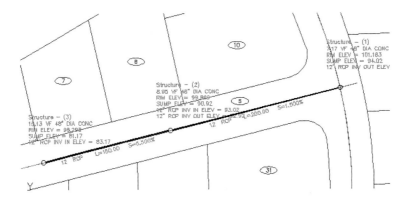

Pipe Networks Not Requiring Structures

You can use Pipes Only mode to create models of entire systems that use only pipes. For example, you can design a water distribution network or a network of conduits that are not pipes, such as electrical lines and fiber optic casings. For a water distribution network, you can draft a design for the network, but the pipe network objects do not model the function of the network.

Creating Pipe Networks

The following procedures show you how to create pipe networks. When you create a pipe network, you specify a default parts list that controls which parts you can create, and the surface and alignment data that is referenced as you create network parts. You then add parts to the network using a pipe network part creation mode.

Procedure: Creating a Pipe Network

The following steps describe how to create a pipe network with a default configuration.

1   On the ribbon, Home tab, Create Design panel, click Pipe Network > Pipe Network Creation Tools. The Create Pipe Network dialog box is displayed.

2   Under Network Name, enter a name for your network.

3   From the Network Parts list, select a parts list that includes the pipes and structures that you want to create.

4   From the Surface Name list, select the default surface that should determine the vertical position of network parts.

5   From the Alignment Name list, select the default alignment to use as a source of stationing data for your pipe network labels.

6　From the Structure Label Style list, select the label style to add automatically to structures as they are created.

7　From the Pipe Label Style list, select the label style to add automatically to pipes as they are created.

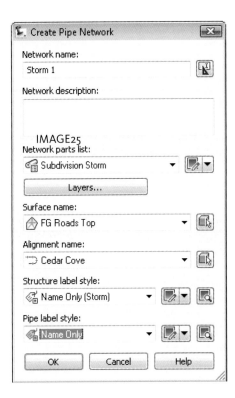

Procedure: Creating Parts in a Pipe Network

The following steps describe how to create parts for a pipe network using the Pipes and Structures pipe network part creation mode. You create parts by selecting locations for structures, which are then connected by pipes.

1　On the Network Layout Tools toolbar, from the Structures list, select the type of structure to create.

2　From the Pipes list, select the type of pipe to use to connect the structures.

3   Set the Toggle Upslope/Downslope button to create pipes that travel in the
     required direction.

4   From the list of pipe network part creation modes, select Pipes and Structures.

5   In the drawing area, click the location for the first structure.

     A structure is created at the location. The vertical placement of the structure is determined
     from the elevation data of the referenced surface.

6   Click the location for a second structure.

     A structure is created at the second location. A pipe is created that connects the first and
     second structures. The elevation and grade of the pipe are determined using the referenced
     surface and the design rules for the selected pipe type.

7   Add additional structures as required. You can change the type of structure and pipes that are
     created as you continue with your layout.

Procedure: Creating a Pipes Only Network

The following steps describe how to use the Pipes Only pipe network part creation mode.

1   In the drawing area, click a pipe network part.

2   On the Network Layout Tools toolbar, from the Pipes list, select the type of pipe to use to
     connect the structures.

3   From the pipe network part creation modes list, select Pipes Only.

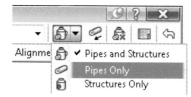

4   Click the location for the start point, then the endpoint of the pipe.

The pipe is created. The endpoint of the current pipe is the start point for the next pipe you draw.

5   Click the location for the endpoint of the second pipe.

The first and second pipe are joined with a null structure.

6   To select a start point that is disconnected from the previous pipe, on the command line, enter **s**.

Click the start point for the next pipe.

Procedure: Creating a Structures Only Network

The following steps describe how to use the Structures Only pipe network part creation mode.

1   In the drawing area, click a pipe network part.

2   On the Network Layout Tools toolbar, from the Structures list, select the type of structure to create.

3   From the pipe network part creation modes list, select Structures Only.

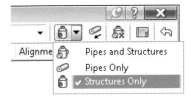

4   In the drawing area, click the locations for structures, as required.

Guidelines for Creating Pipe Networks

Keep the following guidelines in mind when you create pipe networks.

- When you create a pipe network, use the Station and Offset transparent command on the Transparent Commands toolbar to create structures based on a fixed offset from an alignment.

- When you create the drawing template for your organization, include parts lists that contain the pipe and structure parts your team would use on a regular basis.

- You can automatically check for interferences between multiple pipe networks using Pipes menu > Utilities > Create Interference Check.

- To keep drawings free of annotation, create tables that show pipe network data.

Example

The following illustrations show the development of pipe networks using two pipe network part creation modes.

In the following illustration, structures have been created at regular intervals along the alignment without pipes using the Structures Only pipe network part creation mode.

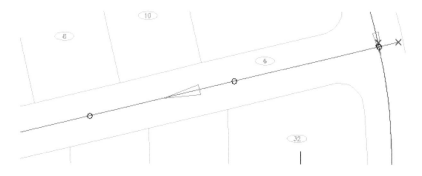

In the next illustration, a second pipe network is created to show the placement of culverts in the project. This network is made up of single pipes created using the Pipes Only pipe network part creation mode.

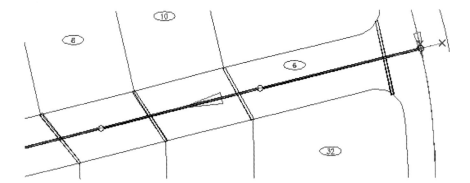

Exercise 01 | Create a Pipe Network

In this exercise, you use the Network Layout Tools to create a storm sewer pipe network for Cedar Cove in plan view. To assist with the creation of the pipe network, you use the Station and Offset transparent command to accurately position the structures adjacent to the alignment.

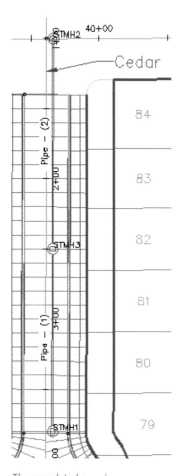

The completed exercise

1   Open...\Site Design - Pipes\I_create_pipe_network.dwg (M_create_pipe_network.dwg).

You use the Transparent Commands toolbar to help layout the pipe network.

2   If the Transparent Commands toolbar is not displayed, on the command line:

- Enter **–toolbar**. Press ENTER.

- Enter **transparent_commands**. Press ENTER.

- Press ENTER.

The Transparent Commands toolbar is now visible.

3   In the drawing area, click the surface.

4   On the contextual ribbon, Modify panel, click Surface Properties.

5   In the Surface Properties - FG Roads Top dialog box, Information tab, for Surface Style, click _No Display. Click OK.

Next, you create the pipe network in plan view.

6   On the ribbon, Home tab, Create Design panel, click Pipe Network > Pipe Network Creation Tools.

7   In the Create Pipe Network dialog box:

- For Network Name, enter **Storm 1**.

- For Network Parts List, click Subdivision Storm.

- For Surface Name, click FG Roads Top. This is the surface you turned off.

- For Alignment Name, click <none>.

- For Structure Label Style, click Name Only (Storm).

- For Pipe Label Style, click Name Only.

- Click OK.

The Network Layout Tools toolbar is displayed.

Next, you select the structure type and pipe size to use. These can be changed any time when you create the network.

8   On the Network Layout Tools toolbar:

- For Structure List, click Concentric Structure 48 diameter
  (Concentric Structure 1,200 diameter).

- For Pipe List, click 15 inch Concrete Pipe (300 mm Concrete Pipe).

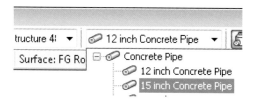

- Ensure the Upslope/Downslope toggle is set to Downslope. You draw the pipe network
  from upslope to down slope.

9   Zoom to the Cedar Cove alignment. This is the eastern cul-de-sac.

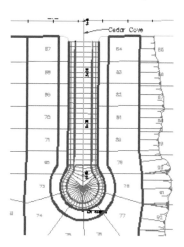

10 On the Network Layout Tools toolbar, click Pipes and Structures. You are prompted for the Structure Insertion Point.

Next, you use the Transparent Commands to locate the structure based on a station and offset from the Cedar Cove alignment. If you make a mistake, you can click Undo on the Network Layout Tools toolbar. If the Network Layout Tools toolbar is closed, click the pipe network (in the drawing area, or in Prospector). Right-click, click Edit.

11 You are prompted to Specify the Structure Insertion Point. On the Transparent Commands toolbar, click Station Offset.

12 When prompted to Select the Alignment, select the Cedar Cove centerline alignment.

13 Move your mouse up and down the Cedar Cove alignment. Note the stations displayed relative to the alignment.

14 When prompted to Specify Station along Alignment, enter **380 (1085** m). Press ENTER.

15 When prompted to Specify Station Offset, enter **-5 (-2 m)**. Press ENTER.

The first structure is created.

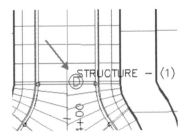

Next, you specify the location for the next structure.

16 When prompted to specify Station Along Alignment, enter **250 (1050** m). Press ENTER.

17 When prompted to Specify Station Offset, enter **-5 (-2** m). Press ENTER.

The next structure is created and connected to the first structure with a pipe.

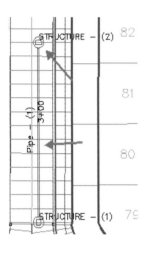

Next, you specify the location for the last structure.

18 When prompted to Specify Station along Alignment, enter **100 (1000** m). Press ENTER.

19 When prompted to Specify Station Offset, enter **-5 (-2** m). Press ENTER.

The last structure is created and a second pipe is added.

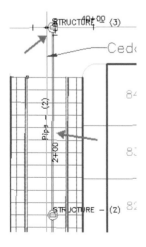

20   Close the Network Layout Tools toolbar.

Next, you examine the pipe network in the Toolspace window.

21   In Prospector, expand Pipe Networks, Networks, Storm 1. Click Pipes.

The pipe data displays in the Item View area. You can edit the data values for the pipe network in this area. You can right-click any column header to control which data columns to view.

22   Click Structures.

23   In the Item View area, Name column, rename the three structures to **STMH1**, **STMH2**, and **STMH3**. You may need to click the Name column header to sort the list.

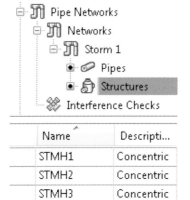

In the drawing, the structure labels update.

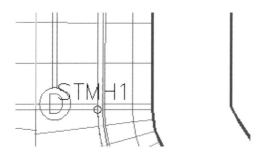

24 The completed drawing appears as follows:

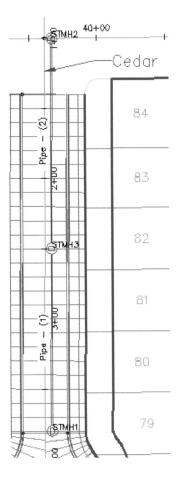

25 Close the drawing. Do not save the changes.

Lesson 35 | Drawing and Editing Pipe Networks

This lesson describes how you draw pipe network parts in profile views and how you edit the pipe network.

When you draw pipe network parts in profile view, you can evaluate the engineering attributes of your design. You can also customize the appearance of labels to help you design, or meet internal or client CAD standards requirements. When you edit pipe network data, the pipe network objects and labels in plan and profile view automatically update to reflect your revisions. This makes it very easy to generate and evaluate alternatives during the planning and detailed design processes. Furthermore, when you edit the plan view location of pipe network parts, the pipe network parts in the profile view automatically update.

The following illustration shows a pipe network in a profile view.

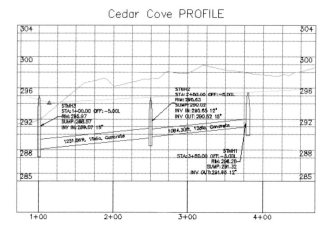

Objectives

After completing this lesson, you will be able to:

- Describe how you draw pipe networks in plan and profile view.

- Describe how you edit pipe networks graphically and in tables.

- List the guidelines for drawing and editing pipe networks.

- Draw a pipe network in profile view.

- Edit the pipe network by moving a manhole and changing a pipe size.

Drawing Pipe Networks in Profile View

When you create a pipe network, you always draw the pipe network in the plan view first. As a separate command, you can then draw the pipes in profile view.

Drawing Pipe Networks in Profile View

When you create a pipe network, you always draw the pipe network in the plan view by positioning pipe network structures and connected pipes. The initial pipe invert elevations and pipe slopes are calculated using pipe and structure rules. After you draw the pipe network in the plan view, you use the Draw Parts in Profile command to draw the pipe network parts in profile view.

You can either draw individual network parts or the entire pipe network in profile view. You can draw pipe network parts in any profile view. This is useful when you want to show crossing pipes for intersecting alignments.

Editing Pipe Networks

You can edit a pipe network either graphically or by changing the pipe data in a table.

Graphical Edits

To edit a pipe network graphically in plan view, select the pipe network part in the drawing area to activate the grips. Pipe network structures and pipes each have their own grips. When you edit a pipe network graphically, the tabular data is automatically updated.

Plan View Grips

Pipe structure grips for plan view graphical editing are shown in the following illustration:

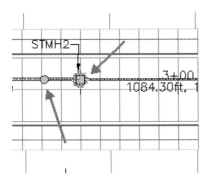

You use the circular grip to rotate the structure. This is useful for asymmetrical structures and structure styles that display text. You use the square grip to change the location of the structure.

When you move a structure, connected pipes move with the structure. Profile structures and pipes locations and associated annotation automatically updates.

 When you use grips to change the location of pipes in plan view, you disconnect the structure from the pipe.

Profile View Grips

There are similar grips that can be used to graphically edit structures and pipes in the profile view. These are shown in the following illustration:

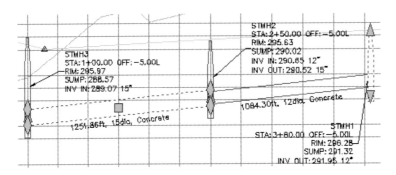

You use diamond-shaped grips on pipe parts to change the invert and obvert elevation for each end of the pipe. This results in a pipe grade change.

You use the square grip on the pipe part to change the invert and obvert elevations at both ends of the pipe. This maintains the pipe grade.

You use the triangle shape grips on the structure pipes to change the rim and sump elevations.

Data Table Edits

There are several options for editing pipe network data in a table. When you edit pipe network data in a table, the graphical display of the pipe and structure objects, and associated annotation, automatically updates.

Prospector Table

You can edit pipe and structure data on the Prospector tab of the Toolspace window.

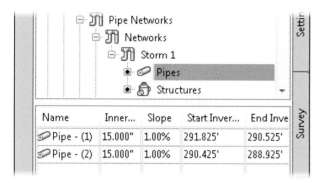

Pipe Properties

You can also edit the Pipe Properties or Structure Properties. Each dialog box displays the engineering properties of the pipe or structure. The Pipe Properties dialog box is shown in the following illustration:

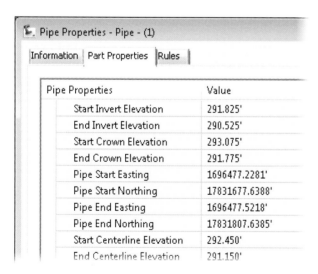

Panorama

You can open the Panorama window to edit both pipe and structure parts. You can also set and preconfigure data column configuration to show the pipe network data you need. The Panorama window is shown in the following illustration:

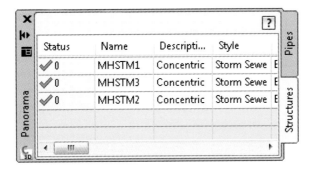

Guidelines for Drawing and Editing Pipe Networks

Keep the following guidelines in mind when you draw and edit pipe networks.

- A single pipe style controls the display of pipe parts in plan, profile view, and section view. A single structure style controls the display of structure parts in plan, profile view, section view, and 3D views.

- Pipe styles and structure styles should be developed and saved in your company/client DWT drawing template.

- When you edit pipe network data in Prospector and Panorama, you can control and pre-configure the data columns to display.

- To provide additional engineering details for the construction of the pipe network, draw pipes and structures in profile view.

Exercise 01 | Draw Pipes in Profile View

In this exercise, you draw the pipe network in the profile view.

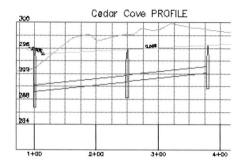

The completed exercise

1   Open \Site Design - Pipes\I_pipes_in_profile.dwg (M_pipes_in_profile.dwg).

First, you split the screen into two views.

2   On the command line, enter **VPORTS**. Press ENTER.

3   In the Viewports dialog box, click Two: Vertical. Click OK.

The screen splits into two vertical views.

4   In the drawing area, click in the left viewport.

5   In Prospector:

   • Expand Pipe Networks, Networks.

   • Right-click Storm 1. Click Zoom To.

The drawing zooms to the plan view for the Storm 1 pipe network in the left viewport.

6   In the drawing area, click in the right viewport.

7 In Prospector:

- Expand Alignments, Centerline Alignments, Cedar Cove, and Profile Views.

- Right-click Cedar Cove PV1. Click Zoom To. The drawing zooms to the Cedar Cove profile view in the right viewport.

Next, you draw pipes in the profile view.

8 Click in the left viewport.

9 On the ribbon, Modify tab, Design panel, click Pipe Network.

10 On the Pipe Network contextual ribbon, Network Tools panel, click Draw Parts in Profile.

11 When prompted to Select Network(s) to Add to Profile View, select any part of the pipe network. Press ENTER.

12 When prompted to Select the Profile View, click in the right viewport. Click the Cedar Cove profile view.

The profile view expands and the pipes are drawn.

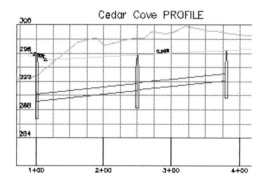

13 Close the drawing. Do not save the changes.

Exercise 02 | Edit a Pipe Network

In this exercise, you edit the pipe network by moving a manhole and changing a pipe size. These edits result in automatic updates to the pipes in the profile view.

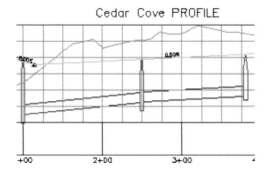

The completed exercise

1   Open \Site Design - Pipes\I_edit_pipes.dwg (M_edit_pipes.dwg).

2   Notice the sump depths for the manholes are too deep.

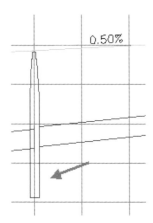

   Next, you modify the sump depths.

3   In the right viewport, select the second manhole (MHSTM2).

4   On the contextual ribbon tab, Modify panel, click Structure Properties.

5   In the Structure Properties dialog box, Part Properties tab, Under Sump Behavior, for Sump Depth, enter **0.5' (0.15 m)**. Click OK.

6   Press ESC.

7   Repeat steps 5 and 6 to modify the sump depths for the other two manholes.

Next, you adjust the position of the manholes in plan view.

8   In the left viewport, select the north pipe segment.

9   Select the square grip at the north end of the pipe segment.

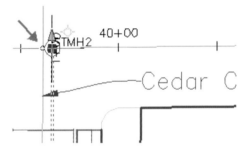

10   Move the pipe to a new location.

Note that the profile view updates to reflect the new pipe location. Note that the structure location does not change and that the pipe is separated from the structure.

11   Click Undo. The pipe is restored to its original location and the profile view is updated.

Next, you move a manhole.

12   Select the north manhole. Click the square grip. Move the manhole below and to the left of the original location.

The manhole and the connected pipe move. The profile updates to reflect the change.

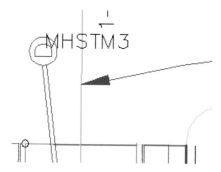

13   Click Undo.

Next, you change the pipe size for one of the pipes.

14   In the plan view, select the south pipe segment. Right-click the pipe segment. Click Swap Part.

15   In the Swap Part Size dialog box, click 12 inch Concrete Pipe (250 mm Concrete Pipe). Click OK.

The part is swapped and the profile view updates to reflect the change. Next, you review the Pipe Rules.

16   In Toolspace, Settings tab, expand Pipe, Pipe Rule Set. Double-click Basic.

17   In the Pipe Rule Set - Basic dialog box, Rules tab:

- Expand Cover and Slope.

- Note that the rule sets the minimum Slope to 1% and the minimum cover to 3' (3m). This rule was applied when the pipe network was created.

- Click Cancel.

18   In Prospector:

- Expand Pipe Networks, Networks, Storm 1.

- Click Pipes.

- In the item view area, notice that the pipes slope at 1% (You may need to scroll to the right to notice the slope columns).

19   In the Item View area, right-click Pipe - (1). Click Pipe Properties.

20   In the Pipe Properties - Pipe - (1) dialog box, Part Properties tab, under Geometry:

- For Pipe Slope (Hold End), enter **0.5%**.

- Click Apply.

21   Click the Rules tab.

- Note the minimum slope violation. In this instance, the engineer decides that the rules violation is acceptable. The minimum allowable pipe slope is actually 0.5%.

- Click OK.

In the drawing area, the slope of the first pipe is adjusted. The final drawing appears as follows:

IMAGE29

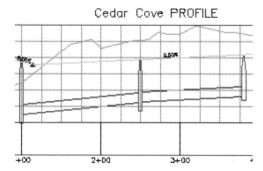

Lesson 36 | Labeling Pipes

This lesson describes how you label pipe networks in plan and profile view.

When you label a pipe network, you display the engineering data that you need to complete and evaluate the design and to construct the pipe network. Pipe labels can be created when you create the pipe network or after you create the pipe network. Pipe labels automatically update when you make changes to the pipe network.

The following illustration shows a labeled pipe network.

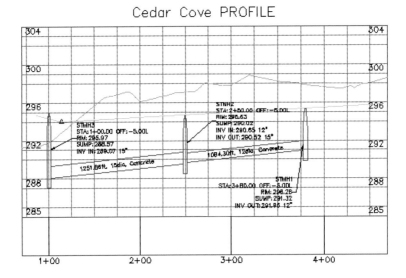

Objectives

After completing this lesson, you will be able to:

- List the components of pipes that you label.

- Describe how you label pipes in plan and profile view.

- Label pipe networks.

Pipe Labels

You create pipe labels to convey engineering and design information.

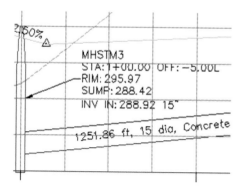

Pipe Labels

You can label any property of a pipe network, structure, or pipe, in plan or profile view. In plan view, you typically label manhole and catch basin identification numbers. For pipes you typically label the length, description (type), and slope. In profile views, you typically label the same information, as well as the invert elevations at the structure locations and rim elevations.

Labeling Pipes

You create pipe labels when you create the pipe network, or after you create the pipe network. Pipe labels automatically update when you make changes to the pipe network.

Labeling Pipes

You use pipe label styles to label pipe network pipes, and you use structure label styles to label pipe network structures. Pipe and structure label styles are found in the Settings tab of Toolspace. This is shown in the following illustration:

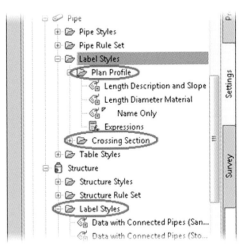

If you need to show different pipe and structure data in plan and profile, you create separate pipe and structure labels for plan and profile views. You can also create pipe label styles to label crossing pipes in the profile view.

Guidelines

Keep the following guidelines in mind when labeling pipes.

- Modify the command settings for pipe networks to set the default pipe and structure label styles. When you do this, the correct label styles are automatically applied when you label the pipe network.

- Use spanning labels to label lengths and slopes over multiple pipe segments. Spanning labels are useful when you want to label the length of an entire pipe network that spans several structures, or if you want to label a pipe network with null structures, such as a watermain network.

Exercise 01 | Label Pipes

Pipe networks are labeled with structure label styles and pipe label styles. In this exercise, you create and apply label styles to your pipe network in both plan and profile views.

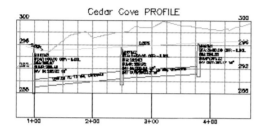

The completed exercise

1   Open *\Site Design - Pipes\I_label_pipes.dwg (M_label_pipes.dwg)*.

You begin by creating the labels for the pipes in plan view. You create and apply a label that shows pipe length, diameter, and material.

2   In Toolspace, Settings tab:

   • Expand Pipe, Label Styles, Plan Profile.

   • Right-click Name Only. Click Copy.

3   In the Label Style Composer dialog box, Information tab, for Name enter Length Diameter Material.

Next, you create a new text component called Pipe Data.

4   On the Layout tab:

   • Click Delete Component.

- Click Create Text Component > Text.

- Under General, for Name, enter **Pipe Data**.

- Under Text, for Contents, click in the Value column. Click the ellipsis.

The Text Component Editor displays.

5   Widen the Text Component Editor dialog box.

6   In the Text Component Editor:

- In the preview area, delete "Label Text."

- Under Properties, click 2D Length – To Inside Edges.

- For Precision, select 0.01.

- Click the right arrow to add the property.

The property, with the formatting, is added to the Text Component editor.

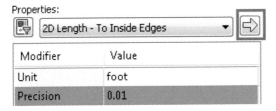

- In the preview area, at the end of the existing text, enter **ft**, (**m**,). That is, enter ft followed by a comma and a space (m followed by a comma and a space).

7   Specify the next property:

- Under Properties, click Inner Pipe Diameter.

- For Precision, select 1.

- Click the right arrow to add the property.

- In the preview area, at the end of the existing text, enter **dia**, Concrete. That is, enter a space, then dia followed by a comma and a space, and then the word Concrete, then a space.

- Click OK.

8   In the Label Style Composer dialog box, under Text, for Y Offset. enter **.1 (2** mm**)**. Press ENTER. Click OK.

Next, you remove the existing pipe labels and apply the new pipe labels to plan view.

9   In the left viewport:

- Select a pipe label in plan view.

- Right-click the pipe label. Click Select Similar.

- Right-click, select Label Properties.

- In AutoCAD Properties, for Pipe Label Style, click Length Diameter Material.

10   The pipes are labeled with the new label style.

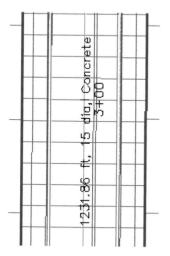

11   Press ESC to end the command.

12   Select the south manhole label.

13   Click the diamond grip. Move the label away from the manhole.

14   Repeat the steps for the other manhole labels.

      The plan view labeling is complete. Next, you label the profile views.

15   Click in the right viewport.

16   On the ribbon, Annotate tab, Label & Tables panel, click the tag on Add Labels.

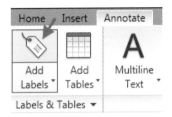

17   In the Add Labels dialog box:

      • For Feature, click Pipe Network.

      • For Label Type, click Entire Network Profile.

      • For Pipe Label Style, click Length Diameter Material.

      • For Structure Label Style, click Data with Connected Pipes (Storm).

      • Click Add.

18   When prompted to Select Part, select any part of the pipe network in the profile view.

      The pipe network parts are labeled in the profile view. The finished drawing
      appears as follows:

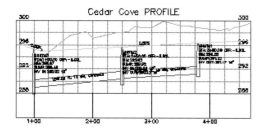

19 Close the drawing. Do not save the changes.

Lesson 37 | Designing Storm Sewer Networks

This lesson describes how you calculate pipe sizes and invert elevations for a storm sewer pipe network, using the Hydraflow Storm Sewers Extension.

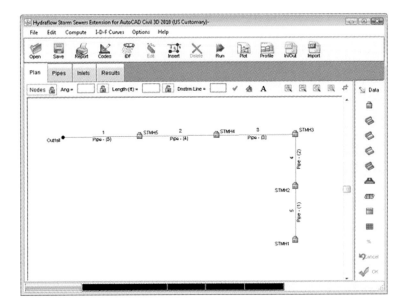

Objectives

After completing this lesson, you will be able to:

- Describe methods for designing a storm sewer network.

- Design a storm sewer network.

- Design a storm sewer network that includes pipe sizes and invert elevations.

About Storm Sewer Networks

This section describes storm sewer networks and the Hydraflow Storm Sewers Extension. You use the Hydraflow Storm Sewers Extension to analyze and calculate storm sewer pipe networks to ensure pipe diameters and invert elevation support designated flow rates.

Definition of Storm Sewer Networks

Storm sewer networks are a series of connected catch basins, manholes, and pipes used to discharge storm water to an outfall location. Pipe diameters and invert elevations in a storm sewer network are calculated based on hydrologic and hydraulic analysis.

Design Calculations

The design of a storm sewer network involves the calculation of pipe diameters and rim elevations from inputted flow data. When you create a pipe network in Civil 3D, you individually select pipe sizes from a list. Invert elevations are typically calculated based on minimum depth of cover and drop across structure rules. However, to meet the requirements of a storm sewer, network pipes must be resized and invert elevations must be recalculated using appropriate hydrologic and hydraulic analysis.

You can either manually enter flow data or use Hydraflow Storm Sewers Extension to calculate the values using traditional methods.

Hydraflow Storm Sewers Extension

The Hydraflow Storm Sewers Extension:

- Is a Civil 3D extension that can read pipe network geometry, pipe types, and structure types created in a Civil 3D pipe network.

- Performs hydraulic analysis of both simple and complex storm sewer networks.

- Can calculate pipe diameters, invert elevations, and energy grade lines for up to 250 connected storm sewer lines.

Examples

The following illustrations show a storm sewer network in Hydraflow Storm Sewers Extension in several views.

Storm Sewer Network In Plan View

The following illustration shows the layout of a storm sewer network in plan view.

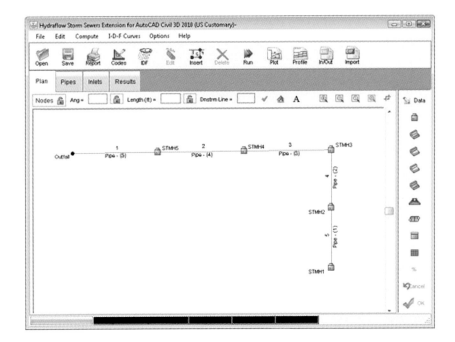

Storm Sewer Network In Profile View

The following illustration shows a storm sewer network in profile view.

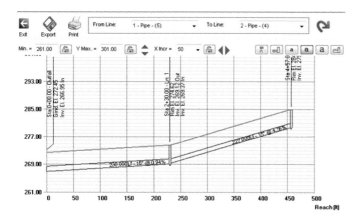

Pipe Data

The following illustration shows the pipe data for the same storm sewer network. You can either input the flow data manually, or you can use Hydraflow Hydrographs Extension to calculate the surface runoff, and resultant flow to the individual pipes in the pipe network.

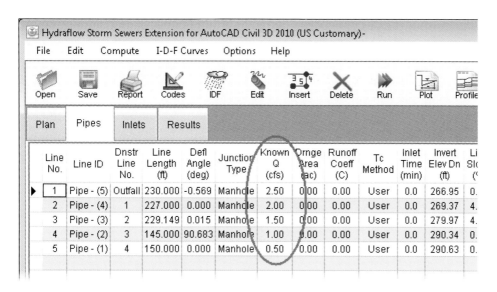

Designing Storm Sewer Networks

This section describes a process for laying out and designing storm sewer networks using Civil 3D and Hydraflow Storm Sewers Extension. When you design a storm sewer network, you layout the pipe network in Civil 3D, export the pipe network to Hydraflow Storm Sewer Extension to calculate flow values for the network, and import the Hydraflow Storm Sewers Extension pipe network to Civil 3D.

Process: Designing a Storm Sewer Network

Laying out and designing an storm sewer network involves working in both AutoCAD® Civil 3D® and Hydraflow Storm Sewers Extension. To layout and design a storm sewer network, follow these steps:

1   Layout the pipe network in Civil 3D using the Pipe Network Creation Tools.

2   Export the pipe network to a Hydraflow Storm Sewers Extension project file.

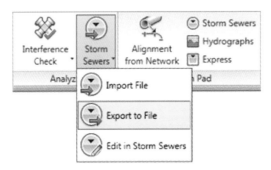

3   Start Hydraflow Storm Sewers Extension, open the project file you exported from Civil 3D, and input the flow values for each pipe.

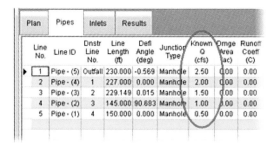

4 Compute the pipe sizes, invert elevations and hydraulic/energy grade lines.

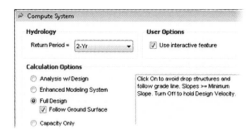

5 Export the designed pipe network to a Hydraflow Storm Sewers Extension project file.

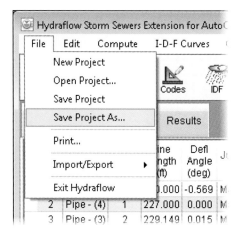

6   Import the pipe network to Civil 3D from the Hydraflow Storm Sewers Extension project file.

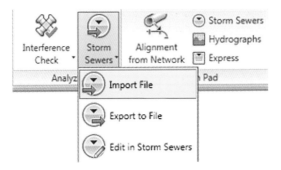

7   Update the storm sewer network with the new data.

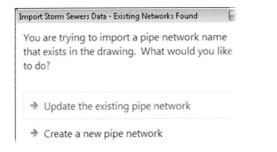

Guidelines

Keep the following guidelines in mind when you create pipe networks.

- Assign the Manning n (smoothness coefficient) value to pipes in the parts list in Civil 3D. When you create a pipe network from parts in the parts list, the Manning n value is transferred to Hydraflow Storm Sewers Extension. Otherwise you need to assign the Manning n value manually in Hydraflow Storm Sewers Extension.

- In Hydraflow Storm Sewers extension, you must specify US Customary for imperial units or SI for metric units.

- In Hydraflow Storm Sewers Extension ensure your design codes are in accordance with local standards for pipe sizing.

Exercise 01 | Design a Storm Sewer Network

In this exercise, you export a basic pipe network to Hydraflow Storm Sewer Extension, compute the storm sewer network values, and import the data to Civil 3D.

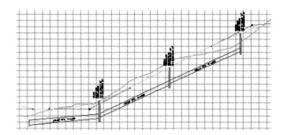

The completed exercise

1   Open *Site Design - Pipes\I_design_pipes.dwg* (*M_design_pipes.dwg*).

The drawing contains a pipe network that was created using the AutoCAD Civil 3D Pipe Network Creation Tools. Review the pipe network in profile view and note that all pipe diameters are the same. You now export the pipe network to a Hydraflow Storm Sewers Extension project file.

2   In the drawing area, select any part of the storm sewer network in plan view.

3   On the contextual ribbon, Analyze panel, click Storm Sewers > Export to File.

4   In the Export to Storm Sewers dialog box, click OK.

5   In the Export Storm Sewers to File dialog box:

   • Browse to ...*Site Design - Pipes*.

   • For File Name, enter **Storm 1 Layout**.

   • Click Save.

6   On the contextual ribbon, Launch Pad panel, click Storm Sewers.

The Hydraflow Storm Sewers Extension launches.

7   If you are working in metric units, you need to change the units in the Hydraflow Extension. Click Options menu > Units > SI.

8   In Hydraflow Storm Sewers Extension, on the toolbar, click Open.

Note: To see the line ID labels, from the menu, click Options > Plan View > Labels > Show Line Ids.

9   In the Open Project dialog box:

- Browse to ...*Site Design – Pipes*.

- Click Storm 1 Layout.stm. Click Open.

Hydraflow Storm Sewers Extension displays the pipe network in plan view.

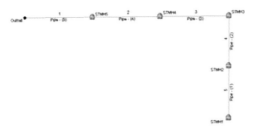

Note: Click Options menu > Plan View > Labels > Show Line Ids.

Next, you enter/confirm the Design Codes, which are the design parameters used for the calculation of pipe sizes and invert elevations.

10   On the toolbar, click Codes.

Report   Codes   IDF

11   In the Design Codes dialog box, Pipes tab, set the following values:

- Minimum Pipe Size, **12** in **(75** mm**)**

- Maximum Pipe Size, **102** in **(3000** mm**)**

- Design Velocity, **2** ft/s **(.75** m/s**)**

- Minimum Slope, **0.5**%

- Maximum Slope, **10**%

- Minimum Cover, **4** ft **(1.5** m**)**

- Default n-value, **0.013**

- For Alignment, check Match Inverts

- Matchline drop, **0.1** ft **(0.03** m**)**

12   Click OK.

13   Click the Pipes tab and In the Known Q column, starting in the Pipe – (1) row, enter:

- Pipe – (1): **0.5** cfs **(0.015** cms**)**.

- Pipe – (2): **1.0** cfs **(0.03** cms**)**.

- Pipe – (3): **1.5** cfs **(0.045** cms**)**.

- Pipe – (4): **2.0** cfs **(0.06** cms**)**.

- Pipe – (5): **2.5** cfs **(0.075** cms**)**.

Line ID	Dnstr Line No.	Line Length (ft)	Defl Angle (deg)	Junction Type	Known Q (cfs)
Pipe - (5)	Outfall	230.000	-0.569	Manhole	2.5
Pipe - (4)	1	227.000	0.000	Manhole	2.0
Pipe - (3)	2	229.149	0.015	Manhole	1.5
Pipe - (2)	3	145.000	90.683	Manhole	1.0
Pipe - (1)	4	150.000	0.000	Manhole	.5

14   Click the green OK check mark icon on the bottom right, to accept the values.

Next, you calculate the pipe sizes.

15   On the toolbar, click Run.

16   In the Compute System dialog box:

- Under Calculation Options, click Full Design.

- Select the Follow Ground Surface check box.

- Under Starting HGLs, in the Starting HGL column, select Normal.

IMAGE35

Invert Elev Dn (ft)	Starting HGL (ft)
265.98	Normal ▼

17   In the Storm Sewers Design dialog box, note that Pipe – (5) is displayed in a profile view. Review the data for Pipe – (5).

18   Click Up to review the data for the remaining pipes in the run.

IMAGE37

20   On the Pipes tab, notice the updated pipe diameters in the Line Span column. The pipes are no longer all 12" diameter and have been resized based on the inputted flow values.

IMAGE39

Invert Elev Up (ft)	Line Rise (in)	Line Type	Line Span (in)	No. Barrels	N Value (n)
268.58	24	Cir	24	1	0.013
279.47	18	Cir	18	1	0.013
289.81	15	Cir	15	1	0.013
290.53	12	Cir	12	1	0.013
291.40	12	Cir	12	1	0.013

view.

21   On the toolbar, click Profile.

22   In the Storm Sewer Profile dialog box, for To Line, select 5 – Pipe (1). Click Update.

Hydraflow Storm Sewers Extension displays the entire pipe run in a profile view.

23   Close the Storm Sewer Profile dialog box.

      Finally, you export design data to a project file, and import to Civil 3D.

24   Click File menu > Save Project As.

25   In the Save Project As dialog box:

      • Browse to ...*Site Design – Pipes*.

      • For File name, enter **Storm 1 Design**. Click Save.

      • Click OK.

26   Close Hydraflow Storm Sewers Extension.

27   Click OK.

      Next, you import the data to Civil 3D.

28   Click in the AutoCAD Civil 3D window.

29   On the contextual ribbon, Analyze panel, click Storm Sewers > Import File.

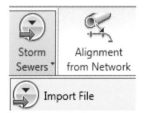

30   In the Import Storm Sewers File dialog box:

      • Browse to ...*Site Design – Pipes*.

      • Click *Storm 1 Design.stm*.

      • Click Open.

31   In the Update Storm Sewers Data dialog box, click Update the Existing Pipe Network.

32 Close Panorama.

The pipe network in Civil 3D is updated with the calculated invert elevations and pipe diameters from Hydraflow Storm Sewers Extension. The pipes in plan and profile view are displayed and annotated with the design data.

The completed exercise drawing appears as follows:

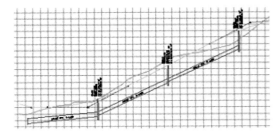

33 Close the drawing. Do not save the changes.

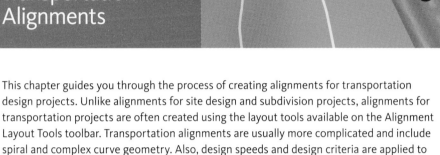

Chapter 11

Transportation - Alignments

This chapter guides you through the process of creating alignments for transportation design projects. Unlike alignments for site design and subdivision projects, alignments for transportation projects are often created using the layout tools available on the Alignment Layout Tools toolbar. Transportation alignments are usually more complicated and include spiral and complex curve geometry. Also, design speeds and design criteria are applied to ensure that curve geometry meets the minimum standards requirements, and to apply superelevation to the alignment.

In the Designing Criteria-Based Alignments lesson, you create an alignment object using tools on the Alignment Layout Tools toolbar, and you modify the properties of the alignment to assign a design speed and design criteria. This helps you identify the substandard curves in the alignment geometry, which are then edited to meet the minimum design standard. This lesson also describes the difference between fixed, floating, and free alignments. In the Applying Superelevation lesson, you apply superelevation to the alignment from data residing in a design criteria file. In the Creating Offset Alignments lesson you learn how to create offset and widening alignments that are geometrically tied and related to a centerline alignment.

▶ Objectives

After completing this chapter, you will be able to:

- Design and edit criteria-based alignments.
- Apply superelevation to alignments.
- Create offset and widening alignments that are geometrically tied and related to a centerline alignment.

Lesson 38 | Designing Criteria-Based Alignments

In this lesson you create a criteria-based alignment and edit it using the layout tools. When engineers plan and design transportation facilities for both new construction and road reconstruction projects, they must create and then edit the alignments used to control the design of the road. You create tangent, curve, and spiral alignment components with layout tools, and you can edit alignments both graphically and in a table.

You interact with the alignment geometry and edit data graphically or directly in a table. When you edit alignment data in a table, the graphical display of the geometry and associated annotation is automatically updated. When you edit alignment data graphically, curves maintain tangency to the lines. When you edit alignment geometry, surface profile data also automatically updates.

The following illustration shows an alignment object.

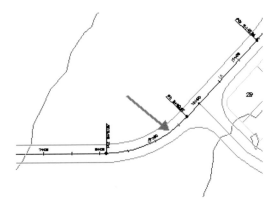

Objectives

After completing this lesson, you will be able to:

- Describe horizontal alignments.

- Describe the options for editing alignments.

- Explain criteria-based design.

- Create and edit a horizontal alignment.

About Alignments

Creating and defining the alignment is one of the first steps in a transportation design project. The alignment controls the horizontal location of a transportation corridor model. When you create an alignment, you can assign design criteria to ensure that you meet the minimum curvature and superelevation requirements for the project.

Definition of Alignments

An alignment is a series of coordinates, lines, curves, or spirals used to represent linear features, such as the centerline of a road, edges of pavement, sidewalks, and rights-of-way.

Entities

Alignments are created and displayed with subentities. The alignment components are lines, curves, spirals, arrows, line extensions, and curve extensions. Alignment lines, curves, and spirals can be either fixed, floating, or free. Fixed, floating, and free entities are summarized in the following table.

Term	Description
Fixed Entities	Alignment lines, curves, and spirals can be fixed entities. Fixed entities have a fixed position and are not necessarily tangent to another entity for the definition of its geometry.
Floating Entities	Alignment lines, curves, and spirals can be floating entities. When a floating entity is created, it is tangent to one other alignment entity for the definition of its geometry.
Free Entities	Alignment lines, curves, and spirals can be free entities. When a free entity is created, it is tangent to two other alignment entities for the definition of its geometry.

Example

An alignment segment with a generated curve is shown in the following illustration.

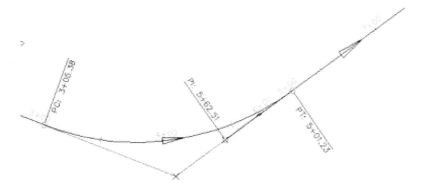

Editing Alignments

You can edit an alignment by grip editing or by changing the layout parameters in a table shown in specialized windows.

Editing with Grips

When you select an alignment, graphical editing grips are displayed. The square grips are used to edit tangents, the circular grips are used to edit curves, and the triangle grips are used to edit the PI location. As you make changes, an alignment retains its tangency rules at curves, and all alignment labels are dynamically updated. The shapes of the grips displayed depend on whether or not the alignment entities are fixed, floating, or free. You can edit the location of tangents and resize curves and spirals.

Grip Type	Description
Circular grip	Changes the parameter of a curve. You can change the radius by moving a center point, pass-through point, or the tangency point of an attached curve. This grip is used only on curves and circles and always affects the radius of the curve.
Square grip	Moves an unconstrained pass-through point on a line, a curve, or a center point of a circle. On a curve or a circle, moving this grip does not affect the radius of the entity that the grip belongs to. However, it may affect the radius of another attached entity.
Triangular grip	Changes where two tangent points meet. This grip is always oriented with the top point up, toward the Y axis of the world coordinate system.

The following illustration shows that when you select an alignment, blue editing grips appear at the curve ends, midpoints, and points of intersection (PIs). By using these grips, you can move alignment features directly and reshape a line or curve using visual cues. Use this method of editing when precision is not important.

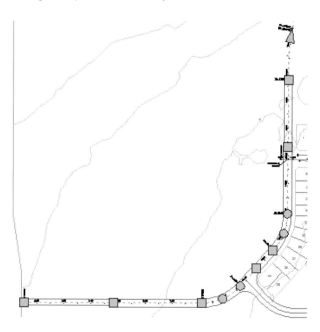

Editing Alignments in Tables

When you edit alignment data in a table, you modify values to change tangent bearings, curve radii, and spiral geometry parameters. The alignment object in the drawing and associated annotation automatically updates. The alignment segments are numbered according to both their position in the alignment and their order of creation. Each row of the table shows design data about a specific alignment entity.

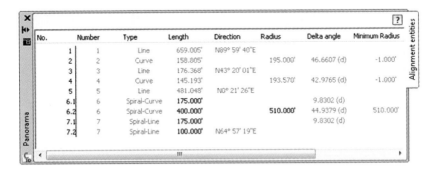

No.	Number	Type	Length	Direction	Radius	Delta angle	Minimum Radius
1	1	Line	659.005'	N89° 59' 40"E			
2	2	Curve	158.805'		195.000'	46.6607 (d)	-1.000'
3	3	Line	176.368'	N43° 20' 01"E			
4	4	Curve	145.193'		193.570'	42.9765 (d)	-1.000'
5	5	Line	481.048'	N0° 21' 26"E			
6.1	6	Spiral-Curve	175.000'			9.8302 (d)	
6.2	6	Spiral-Curve	400.000'		510.000'	44.9379 (d)	510.000'
7.1	7	Spiral-Line	175.000'			9.8302 (d)	
7.2	7	Spiral-Line	100.000'	N64° 57' 19"E			

The design data for an entity that you select in the Alignment Entities window is displayed in the Alignment Layout Parameters window. This simplifies review and edit tasks.

For precise alignment editing, such as when your design calculations and reference tables provide numeric values for minimum curve radius, length, or spiral A values, you use the Alignment Layout Parameters window.

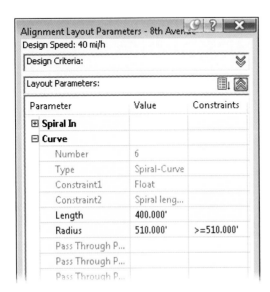

About Criteria-Based Design

Alignments used to define transportation-related entities may be complicated and include spiral and complex curve geometry. The criteria-based design feature provides the ability to verify that your alignment design meets the minimum standards required by your local agency. Options to customize and specify a design criteria file ensures that curve geometry based on design speeds and criteria for your locality are met and the appropriate superelevation is applied to the alignment.

Definition of Criteria-Based Design

Criteria-based design is an option available to the engineer when creating a new alignment or editing an existing one. It enables the engineer to verify that a design meets the local agency's design standards. When this option is used, the alignment is checked against the minimum values specified in a design criteria file. This design criteria file is XML based. The file contains all the local agency's minimum design values for such parameters as the minimum curve radius, superelevation attainment method, and the spiral transition length for a wide range of design speeds.

If this option is used when laying out a new alignment, the appropriate minimum values are displayed on the command line for acceptance. If you apply these design criteria to an existing alignment, then the entire alignment is checked against the design values. In either case, whether the design criteria is applied during alignment layout or applied to an existing alignment after the fact, any segment of the alignment that does not meet the minimum values specified in the design file is flagged with a warning symbol. This warning symbol provides a tooltip with information about the standard that was violated, and what the standard value should be.

Parameters

The following table provides a description of the terms and options available to the engineer when applying Criteria-based design to alignments. Only the horizontal component of an alignment is being described here. The vertical component is discussed in a later lesson.

Term	Definition
Criteria-Based Design	This option enables the use of criteria-based design to be applied to the current alignment
Design Criteria File	This option enables you to specify the location of the XML file that contains the all of the local agency's design standards controlling the minimum horizontal and vertical design parameters.
Default Criteria	This list displays the major design categories and their values that will be used in the layout process. These categories are contained in the design criteria file and include the minimum radius table, the transition length table, and the superelevation attainment method for horizontal layout, as well as the minimum K table used in vertical layout.
Design Checks	This option allows for the use of Design checks against the current alignment. A Design Check is a user-defined expression used to verify that an entity meets the minimum design standards that were established for the alignment or profile object. Design checks may be defined for different entity types, such as lines, curves, and spirals. A design check must be saved in a design check set to be applied to an alignment or profile.

For more information use Help files.

The design criteria file and some criteria values are shown in the following illustration.

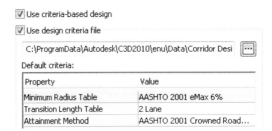

Example

If the design parameters for a subentity violate the minimum values established in the design criteria file, a warning marker displays on the subentity in the drawing window. The display of the warning symbol is controlled by the object's style (profile or alignment).

The following illustration shows a Design Check Marker warning that shows that minimum radius violation has occurred, and that the radius should be a minimum of 60m for the given design speed.

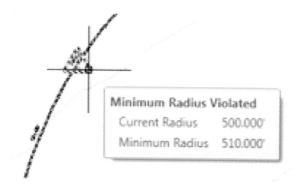

Exercise 01 | Create and Edit an Alignment

In this exercise, you create an alignment using the Alignment Layout Tools toolbar. You then edit the alignment using two different methods.

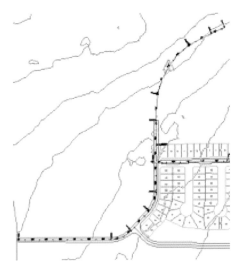

The completed exercise

Create Alignment

First, you create an alignment.

1   Open *Transportation - Alignments\I_alignment_by_layout.dwg*
    (*M_alignment_by_layout.dwg*).

    In the drawing area, notice the collector road alignment along the west side of the proposed subdivision. The south part of the centerline alignment has been designed and is represented with AutoCAD® tangents and arcs.

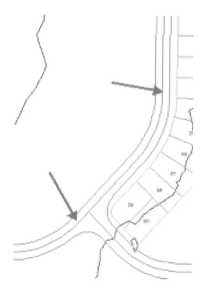

2   On the ribbon, Home tab, Create Design panel, click Alignment › Alignment Creation Tools.

3   In the Create Alignment - Layout dialog box, General tab:

- For Name, enter **8th Avenue**.

- For Type, select Centerline.

- For Site, select ‹None›.

- For Alignment Style, select Layout.

- For Alignment Label Set, select All Labels.

4   Click the Design Criteria tab and:

- For Starting Design Speed, enter **40mi/h (70km/h)**.

- Check Use Design Criteria File check box.

- For Minimum Radius Table, select AASHTO 2001 eMax 6%.

- Clear Use Design Check Set checkbox.

- Click OK.

5   On the Alignment Layout Tools - 8th Avenue toolbar, click Convert AutoCAD Line and Arc.

Spiral Type: Clothoid

6   When prompted to Select Line, Arc to Convert, from left to right, select the five centerline alignment entities (three tangents and two arcs in the order: tangent, arc, tangent, arc, and tangent) that form the south part of 8th Avenue. Press ENTER.

This creates an alignment object from the selected entities. The original line and arc entities are still in the drawing and unnecessarily duplicate the alignment object.

Next, you remove the original line and arc entities.

7   In the drawing area:

   • Select the 8th Avenue alignment object that you created in the previous step.

   • Right-click, click Display Order > Send to Back.

   • Select the three tangents and two arcs you originally selected to create the alignment. Press DELETE.

Next, you construct the remainder of the alignment, which begins with a floating curve attached to the north end of the alignment.

Chapter 11 | Transportation - Alignments

8   On the Alignment Layout Tools - 8th Avenue toolbar, click Floating Curve with Spiral (from Entity End, Radius, Length).

9   When prompted to Select Entity to Attach To, select a point on the tangent near the north end of the alignment.

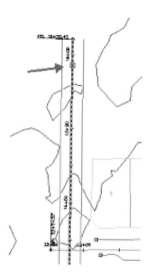

Notice that the default radius is 510' (195 m). This is the minimum value for the assigned design speed.

10   When prompted to:

- Specify Radius, enter **500 (185** m**)**. Press ENTER.

- Specify Spiral in Length, enter **175 (50** m**)**. Press ENTER.

- Specify Curve in Direction, press ENTER to accept Clockwise.

- Specify Length, enter **400 (170** m**)**. Press ENTER twice.

The entrance spiral and the curve of the new portion of the alignment are complete.

Next, you create the exit spiral and tangent.

11 On the Alignment Layout Tools toolbar, click Floating Line with Spiral (From Curve End, Length).

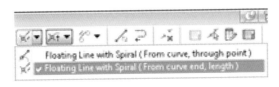

12 When prompted to Select Entity to Attach To, select the last curve near the endpoint of the alignment.

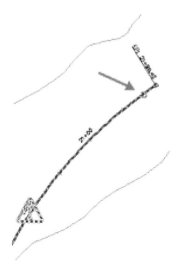

13   When prompted to:

- Specify Spiral in Length, enter **175 (50** m**)**. Press ENTER.

- Specify Line Length, enter **100 (30** m**)**. Press ENTER twice.

This creates the existing spiral and tangent and applies stationing to the remainder of the alignment.

Next, you change the starting station for the design speed. The assigned design speed is only applicable for the new section of the alignment to the north of the subdivision.

14   In the drawing select the alignment. Right-click, click Alignment Properties.

15   In the Alignment Properties - 8th Avenue dialog box, Design Criteria tab:

   • Click in the Start Station cell. Click the Select Station icon.

   • In the drawing area, use the endpoint OSNAP and snap to the northwest corner of
     Parcel number 1.

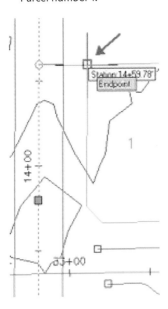

16   Click OK.

     This creates a design speed label in the drawing area. Notice the warning symbol alignment
     component on the curve.

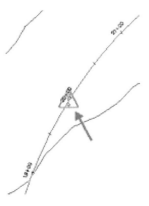

17   On the Alignment Layout Tools toolbar, click Alignment Grid View.

This displays the alignment data in the Panorama window. Notice the minimum radius design criteria violation. You may need to adjust the column display to show the data.

18   Close the Alignment Layout Tools - 8th Avenue toolbar.

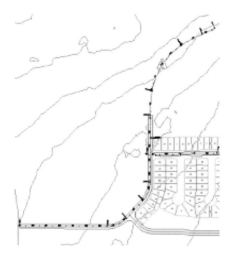

19   Close the drawing. Do not save changes.

Edit Alignment

Next, you edit the alignment.

1   Open \Transportation - Alignments\I_edit_alignment.dwg (M_edit_alignment.dwg).

You begin by editing the alignment data in a table to correct violations to the assigned design criteria. Notice the design criteria warning symbol component at the north end of the alignment.

The path for the design criteria file referenced by the alignment in this drawing is based on Windows Vista. If you are using Windows XP, proceed to step 2. If you are using Windows Vista, proceed to step 3.

2   Set the path for the design criteria file.

- In the drawing area, select the alignment.

- Right-click, click Alignment Properties.

- In the Alignment Properties dialog box, click the Design Criteria tab.

- Under Use Design Criteria File, click the ellipsis.

- In the Select Design Speed Table dialog box, browse to ...\Autodesk\C3D 2010\enu\Data\ Corridor Design Standards\Imperial (Metric).

- Select _Autodesk Civil 3D Imperial (Metric) Roadway Design Standards.xml.

- Click Open.

- For Minimum Radius Table, click AASHTO 2001 eMax 6%.

- Click OK. Press ESC.

3   In the drawing area, select the 8th Avenue alignment. On the contextual ribbon, click
     Geometry Editor.

4   On the Alignment Layout Tools - 8th Avenue toolbar, click Pick Sub-entity. In the drawing area
     select the curve with the design criteria warning symbol.

5   In the Alignment Layout Parameters - 8th Avenue dialog box, notice the design criteria
     violation for radius. Close the Alignment Layout Parameters - 8th Avenue window.

Length	400.000'	
⚠ Radius	500.000'	>=510.000'

6   On the Alignment Layout Tools toolbar, click Alignment Grid View.

7   In Panorama:

     • Notice the design criteria violation for the curve radius.

     • For No. 6.2, click in the Radius column.

     • Change the radius to **510' (195 m)**. Press ENTER.

     The design criteria violation warning symbol is removed in the table and in the drawing, and
     the alignment geometry is updated.

8   Close the Alignment Layout Tools toolbar.

     Next, you edit the alignment graphically.

9   In the drawing area select the 8th Avenue alignment. Zoom to the north end of the
     alignment. Click the grip on the endpoint of the curve.

10   Move the grip to the right.

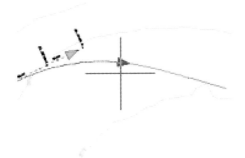

Chapter 11 | Transportation - Alignments

11   Select a location. Click to lengthen the curve.

     The geometry and stationing for the alignment are updated. The existing spiral and tangent
     are floating entities and therefore maintain tangency to the curve.

12   Close the drawing. Do not save changes.

Lesson 39 | Applying Superelevation

This lesson describes how you apply superelevation to alignments. Superelevation is the purposeful canting of road or railway cross section within spiral and curve alignment components. The intent is to counteract centrifugal forces to allow for higher design speeds and safer passage through alignment spirals and curves.

To calculate superelevation values for an alignment, you modify alignment properties to assign a design speed. The superelevation values are referenced from a rate table in a design criteria file. Each curve is a superelevation region, and for each superelevation region you can assign independent superelevation properties. You can also apply the superelevation properties of one superelevation region to the other superelevation regions in the alignment.

When you use alignment design criteria and superelevation, you associate standard imperial and metric superelevation tables with the alignment to calculate minimum curve radii and superelevation critical values for assigned design speeds. The superelevation values assigned to an alignment are referenced when you create the corridor model for the transportation facility.

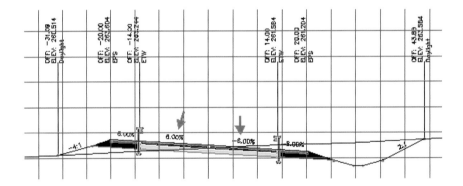

Objectives

After completing this lesson, you will be able to:

- Describe criteria-based design and superelevation.

- List the guidelines for calculating superelevation.

- Calculate and apply superelevation to an horizontal alignment.

About Superelevation

Superelevation is calculated for an alignment when you assign a design speed, a design criteria file, a rate table, and a transition length table. You can assign different design speeds at different locations along the alignment.

Design criteria files are available for both imperial and metric units. The rate table indicates the superelevation rate, or the maximum section cross fall. The transition length table is used to assess the length it takes to achieve full superelevation, and the locations of the critical superelevation points.

The following illustration shows the superelevation specifications for a superelevation region on an alignment. Notice the design criteria file, superelevation rate table, and the transition length table assignment.

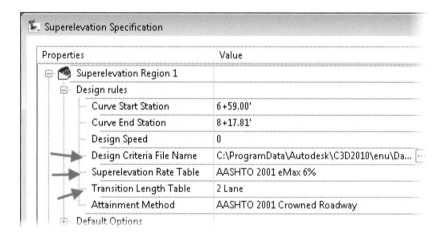

Definition of Superelevation

Superelevation is the purposeful banking of a roadway to provide for smooth and safe travel around a curved section of road at the design speed. The purpose of superelevation is to counteract the centripetal and frictional forces experienced by a vehicle as it moves along this curved path.

Applying Superelevation

Superelevation is defined and applied in the Alignment Properties dialog box, shown below. To calculate superelevation you need to specify the design speed, the minimum radius table, and the superelevation attainment method. The following illustration shows the calculated superelevation critical points for an alignment.

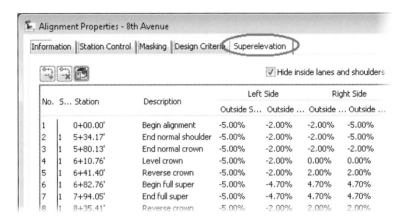

Example

The following illustration shows how superelevation is attained for a clockwise-direction curve on a crowned roadway. To achieve the superelevation, the left edge of pavement rises and the right edge of pavement drops.

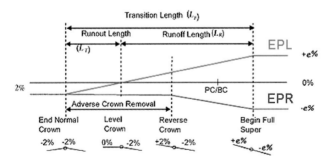

Superelevation is defined and applied in the Alignment Properties dialog box. To calculate superelevation you need to specify the design speed, the minimum radius table, and the superelevation attainment method.

Guidelines for Applying Superelevation

Keep the following guidelines in mind when calculating superelevation for an alignment.

- When assigning superelevation to an alignment, be sure to leave enough tangent between reverse curves to adequately generate sufficient runout and runoff.

- After you apply superelevation to an alignment, you can modify the location of critical superelevation points when you modify the alignment properties.

- If the consecutive curves on an alignment are close together, there may not be sufficient tangent length to transition the superelevation out of one curve and into the next curve. In these instances you can manually adjust the superelevation by adding, removing, and changing the location of superelevation critical points.

Example

The following illustration shows an example of an alignment where the superelevation stations overlap. You can manually adjust the superelevation by adding, removing, and changing the location of superelevation critical points.

No.	S...	Station	Description	Left Side		Right Side	
				Outside S...	Outside ...	Outside ...	Outside ..
1		0+00.00'	Begin alignment	-5.00%	-2.00%	-2.00%	-5.00%
2	1	5+34.17'	End normal shoulder	-5.00%	-2.00%	-2.00%	-5.00%
3	1	5+80.13'	End normal crown	-5.00%	-2.00%	-2.00%	-2.00%
4	1	6+10.76'	Level crown	-5.00%	-2.00%	0.00%	0.00%
5	1	6+41.40'	Reverse crown	-5.00%	-2.00%	2.00%	2.00%
6	1	6+82.76'	Begin full super	-5.00%	-4.70%	4.70%	4.70%
7	1	7+94.05'	End full super	-5.00%	-4.70%	4.70%	4.70%
8	1	8+35.41'	Reverse crown	-5.00%	-2.00%	2.00%	2.00%
9	1	8+66.05'	Level crown	-5.00%	-2.00%	0.00%	0.00%
10	1	8+96.69'	Begin normal crown	-5.00%	-2.00%	-2.00%	-2.00%
11	1	9+42.65'	Begin normal shoulder	-5.00%	-2.00%	-2.00%	-5.00%
12	2	8+68.78'	End normal shoulder	-5.00%	-2.00%	-2.00%	-5.00%
13	2	9+14.74'	End normal crown	-5.00%	-2.00%	-2.00%	-2.00%
14	2	9+45.37'	Level crown	-5.00%	-2.00%	0.00%	0.00%

Exercise 01 | Apply Superelevation

In this exercise, you calculate and apply superelevation to an alignment.

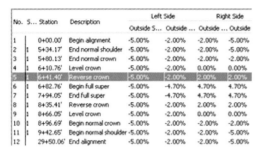

No.	S...	Station	Description	Left Side		Right Side	
				Outside S...	Outside ...	Outside ...	Outside
1	1	0+00.00'	Begin alignment	-5.00%	-2.00%	-2.00%	-5.00%
2	1	5+34.17'	End normal shoulder	-5.00%	-2.00%	-2.00%	-5.00%
3	1	5+80.13'	End normal crown	-5.00%	-2.00%	-2.00%	-2.00%
4	1	6+10.76'	Level crown	-5.00%	-2.00%	0.00%	0.00%
5	1	6+41.40'	Reverse crown	-5.00%	-2.00%	2.00%	2.00%
6	1	6+82.76'	Begin full super	-5.00%	-4.70%	4.70%	4.70%
7	1	7+94.05'	End full super	-5.00%	-4.70%	4.70%	4.70%
8	1	8+35.41'	Reverse crown	-5.00%	-2.00%	2.00%	2.00%
9	1	8+66.05'	Level crown	-5.00%	-2.00%	0.00%	0.00%
10	1	8+96.69'	Begin normal crown	-5.00%	-2.00%	-2.00%	-2.00%
11	1	9+42.65'	Begin normal shoulder	-5.00%	-2.00%	-2.00%	-5.00%
12	1	29+50.06'	End alignment	-5.00%	-2.00%	-2.00%	-5.00%

The completed exercise

1 Open *Transportation - Alignments\I_superelevation.dwg (M_superelevation.dwg)*.

   The path for the design criteria file referenced by the alignment in this drawing is based on using Windows Vista. If you are not using Windows Vista, then proceed to the next step. Otherwise skip the next step and continue to step 3.

2 Set the path for the design criteria file.

   • In the drawing area, select the alignment.

   • On the contextual ribbon, click Alignment Properties.

   • In the Alignment Properties dialog box, click the Design Criteria tab.

   • Under Use Design Criteria File, click the ellipsis.

   • In the Select Design Speed Table dialog box, browse to ...*Autodesk\C3D 2010\enu\Data\ Corridor Design Standards\Imperial (Metric)*.

   • Select *_Autodesk Civil 3D Imperial (Metric) Roadway Design Standards.xml*.

   • Click Open.

   • For Minimum Radius Table, click AASHTO 2001 eMax 6%.

   • Click OK.

3   In the drawing area, select the 8th Avenue alignment. Right-click, click Alignment Properties.

4   On the Alignment Properties - 8th Avenue dialog box, Design Criteria tab, notice the following:

- Design Speed is 40 mi/h (70 km/h).

- Use Criteria-Based Design is selected.

- Use Design Criteria File is selected.

- Minimum Radius Table is AASHTO 2001 eMax 6%.

5   Click the Superelevation tab and Click Set Superelevation Properties to calculate the superelevation.

6   In the Superelevation Specification dialog box:

- Notice the three superelevation regions. Each superelevation region represents a curve in the alignment.

- Collapse Superelevation Region 1 and Superelevation Region 2.

- For Superelevation Region 3, Design Rules, Superelevation Rate Table, select AASHTO 2001 eMax 8%.

- Click OK.

The superelevation for the three curves is calculated. For this exercise, you concentrate on the third curve.

Next, you delete the superelevation data for the first two curves.

7   On the Superelevation tab:

- Scroll to the top of the superelevation data.

- Select the first row.

- Scroll down to the last entry for Superelevation Region 2.

- SHIFT+select line 21. All entries for superelevation regions 1 and 2 (curves 1 and 2) are highlighted.

8   Click Delete Transition Station.

Only the superelevation entries for the third curve remain.

9   Click any cell.

Notice that you can modify superelevation stations, descriptions, and values in this table. You can also add other locations for superelevation.

..	Station	Description	Outside S...
	0+00.00'	Begin alignment	-5.00%
	5+34.17'	End normal shoulder	-5.00%
	5+80.13'	End normal crown	-5.00%
	6+10.76'	Level crown	-5.00%
	6+41.40'	Reverse crown	
	6+82.76'	Begin alignment	
	7+94.05'	Begin normal crown	
	8+35.41'	Begin normal shoulder	
	8+66.05'	Begin full super	
	8+96.69'	End full super	
	9+42.65'	End normal shoulder	
	29+50.06'	End normal crown	
		End alignment	
		Level crown	
		Low shoulder match	
		Reverse crown	
		Shoulder breakover	

- Click OK.

10   Close the drawing. Do not save the changes.

Lesson 40 | Creating Offset Alignments

This lesson describes how you create an offset alignment to model alignment offset features.

Offset alignments are directly related to the centerline alignment. When you change the geometry of the centerline alignment, the geometry of the offset alignment automatically recalculates based on the offset parameters.

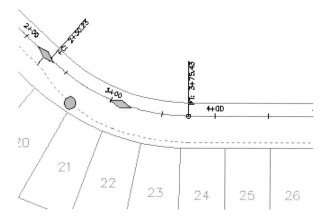

Objectives

After completing this lesson, you will be able to:

- Describe offset alignments.

- Describe the process for creating offset alignments.

- Create offset alignments and widenings.

About Offset Alignments

You use offset alignments to model alignment offset features such as pavement edges, gutter lines and sidewalks.

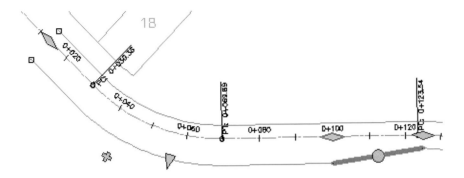

Definition of Offset Alignments

An offset alignment is a dynamic alignment created at an offset distance from another alignment, such as a road edge offset from a centerline alignment. Offset alignments run parallel to the centerline alignment and are defined to model features such as pavement edges, gutter lines and sidewalks.

The geometry of offset alignments are based on the geometry of the centerline alignments. When you edit the geometry of the centerline alignment, the offset alignment will automatically update to follow the geometry of the centerline alignment, at a specified offset value.

Widenings

Widenings expand the width of a roadway for a specified length to accommodate a feature such as a turn lane or bus bay. The widening usually includes a transition region at one or both ends. The Create Widening command simplifies the creation of roadway features such as turn lanes, acceleration lanes, deceleration lanes, and bus bays.

When you create a widening, you can either create a new offset alignment or modify an existing offset alignment.

If the alignment is an offset, it is widened by the value you specified and the widening parameters are added to the properties. If you add a widening to an alignment that is not an offset, the command creates a new dynamic offset alignment. You can edit the widening by using grips or modifying the values in a table.

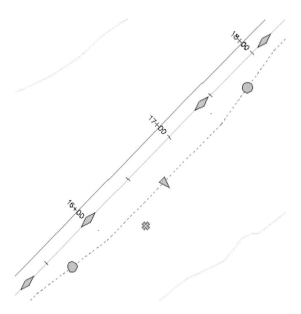

Example

An engineer is planning a two-lane road with parallel lanes located 12' from the left and right sides of the centerline. To accomplish this you would create two offset alignments.

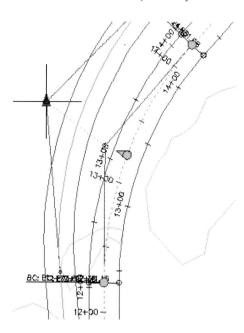

Creating Offset Alignments

To create an offset alignment you, select a centerline alignment (alignment to offset from), specify the starting and ending station of the offset alignment relative to the centerline alignment, and specify the offset distances on the left and right sides. To create multiple offsets, specify how many you want to create for each side.

Process: Creating Offset Alignments

To create an offset alignment, you select: the parent alignment, a station range, and the distance the offset should maintain from the parent alignment. To create multiple offsets, specify how many offsets you want to create from each side.

The following steps show how you create an offset alignment.

1   Launch Create Offset Alignment command. Select alignment location.

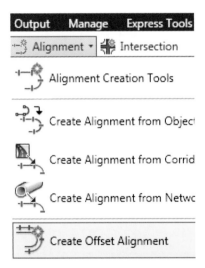

2   Configure offset alignment parameters.

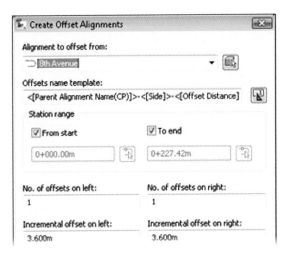

3   Add a widening as required. Enter start station. Enter end station. Enter widening width.

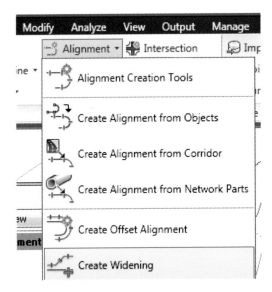

4   Edit widening and transition parameters.

Property	Value
Widening Parameters	
Offset	7.000m
Start Station	0+077.60m
End Station	0+227.42m
Region Length	149.816m
Transition Parameters at Entry	
Transition Type at Entry	Linear
Taper Input Type	By Length
Transition Length	75.000m

Key Points

- For each widening, specify the entry transition length, widening length, and exit transition length.

- Specify default transition lengths in the command settings for the AddWidening command.

- Curb return alignments in intersections can be edited to add widenings at one or both ends. This type of widening forms a turn lane at the entry to a curb return or a merge lane at the exit from a curb return.

- Offset and widening geometry parameters can be modified after they have been created.

Exercise 01 | Create Offset Alignments and Widenings

In this exercise, you create an offset alignment and a widening on an alignment. You also hide the display of an alignment by applying an alignment mask.

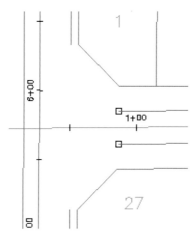

The completed exercise

1   Open *Site Design - Alignments\I_offsets_widening.dwg (M_offsets_widening.dwg)*.

First, you create offset alignments for the 8th Avenue alignment.

2   On the ribbon, Home tab, Create Design panel, click Alignment > Create Offset Alignment.

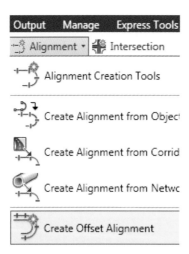

3   When prompted to Select an Alignment, in the drawing area, select the 8th
     Avenue alignment.

     This is the south to north alignment on the west side of the site.

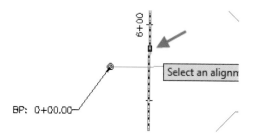

4   In the Create Offset Alignments dialog box:

     • For Incremental Offset on Left, enter **12' (3.6 m)**.

     • For Incremental Offset on Right, enter **12' (3.6 m)**.

     • For Alignment Style, select Proposed Edge of Pavement.

     • For Alignment Label Set, select_No Labels.

     • Click OK.

     Two offset alignments are created.

5   In Prospector, expand Alignments, Offset Alignments. Note the two new offset alignments.

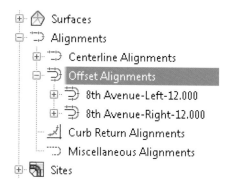

     Next, you create offset alignments for Orchard Road, the east to west alignment inside
     the subdivision.

6   On the ribbon, Home tab, Create Design panel, click Alignment > Create Offset Alignment.

7   When prompted to Select an Alignment, in the drawing area, select Orchard Road.

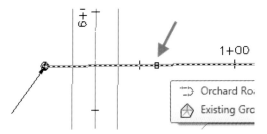

8   In the Create Offset Alignments dialog box:

•   Clear the From Start check box.

•   Click the icon to the right of the start station cell, to pick the starting station in the drawing.

9   When prompted to Specify Station Alignment, snap to the south endpoint of the property line chamfer in parcel number 1.

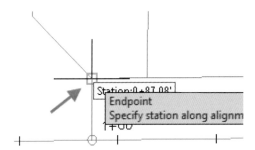

10   In the Create Offset Alignments dialog box:

- For Incremental Offset on Left, enter **12' (3.6 m)**.

- For Incremental Offset on Right, enter **12' (3.6 m)**.

- For Alignment Style, select Proposed Edge of Pavement.

- For Alignment Label Set, select _No Labels.

- Click OK.

Offset alignments are created for Orchard Road.

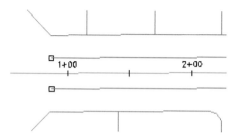

Next, you model a deceleration and acceleration lane for 8th Avenue at the 8th Avenue and
Orchard Road intersection. You add a widening to the right offset alignment.

11   In the drawing area, select the right offset alignment for 8th Avenue.

12   On the contextual ribbon, Modify panel, click Add Widening.

13   When prompted to Create Widening Portion as a New Alignment, click No.

14   When prompted to Select Start Station, enter **300 (100)**. Press ENTER.

15   When prompted to Select End Station, snap to the north endpoint of the 8th
Avenue alignment.

16   When prompted to Enter Widening Offset, enter **24' (7.2 m)**.

The Offset Alignment Parameters palette is displayed.

17   In the Offset Alignment Parameters palette:

- For Transition Length, enter **85' (25 m)**

- Under Transition Parameters at Entry, for Transition Type at Entry, select Curve - Reverse - Curve.

- Close the palette.

18   The offset alignment with an entry transition is modeled in the drawing area. The offset alignment will react when you change the geometry of the 8th Avenue Centerline alignment.

19   In the drawing area, select the 8th Avenue centerline alignment to show the editing grips.

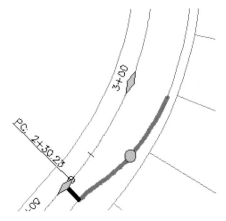

20   Click the blue square grip on the north end of the north tangent.

21   Click a location to the left to relocate the tangent.

The offset alignment with widening changes to follow the 8th Avenue Centerline alignment.

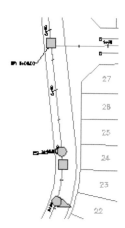

22   On the Quick Access toolbar, click Undo to restore the 8th Avenue to its original location.

Next, you hide a portion of the offset alignment by masking the alignment.

23   In the drawing area, select the 8th Avenue right offset alignment. Right-click, click Alignment Properties.

24   In the Alignment Properties dialog box, click the Masking tab. Click Add Masking Region.

25   When prompted to Specify First Station for Masking Region, select the south chamfer endpoint for the parcel southeast of the 8th Avenue and Orchard Road intersection.

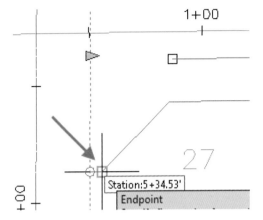

26 When prompted to Specify Second Station for Masking Region, select the north chamfer endpoint for the parcel northeast of the 8th Avenue and Orchard Road intersection.

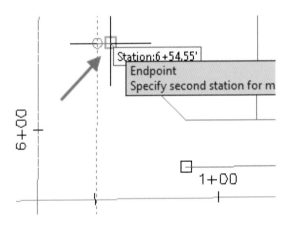

27 In the Alignment Properties dialog box, click OK.

The 8th Avenue right offset alignment is masked in the vicinity of the intersection.

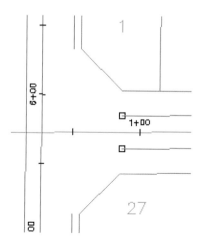

28 Close the drawing. Do not save the changes.

Chapter 12
Transportation - Assemblies and Corridors

In this chapter, you create assemblies and corridors for transportation engineering projects. Assemblies and corridor for transportation engineering projects are usually more complex than those used in site design and subdivision design projects.

Transportation assemblies model a wide variety of typical cross-section configurations including those required for road reconstruction, divided highways, arterials with raised center medians, and interchanges. Furthermore, they often include subassemblies that are used for daylighting to a surface. Corridors are used to model complex transportation facilitates that include many variances in the typical cross-section configuration over different station ranges.

The lessons in the chapter guide you through the process for creating assemblies and corridor models for transportation engineering projects. You create an assembly that models a two-lane rural cross section with daylighting. You also create a corridor model that uses target mapping to model a parallel acceleration lane and a lane taper. You use the Corridor Section Editor to view the design cross- sections at different station locations and apply tabular edits to the cross-section data. Next, you create dynamic top and datum corridor surfaces. In the Intersection Design lesson you learn how to model an intersection and create a corridor model, starting with 2 intersecting alignments and profiles. Finally, you create a 3D model of the proposed design. The 3D model is useful for three-dimensional visualization, and also a powerful asset for public project participation meetings and presentation to government agencies for project approval.

▶ Objectives

After completing this chapter, you will be able to:

- Create and modify transportation assemblies.

- Create corridor models for arterial transportation facilities.

- Create transportation corridor surfaces.

- Create a four-way intersection using the Create Intersection wizard.

- Model road designs in 3D.

Lesson 41 | Creating and Modifying Transportation Assemblies

This lesson describes how you create and modify the properties of assemblies that are used to create transportation corridors that model arterials and freeways.

Assemblies for transportation road and highway corridor models are usually more complex than assemblies used for subdivision road corridor models. An assembly for a transportation corridor model involves the application of alignment superelevation and the use of other alignments and profiles to control how you generate the corridor model. Also, assemblies for transportation corridor models usually incorporate daylighting subassemblies, which project match slopes to a target surface in cut and fill conditions.

The following illustration shows the subassemblies in a transportation corridor assembly that daylights to a surface and reads the alignment superelevation parameters.

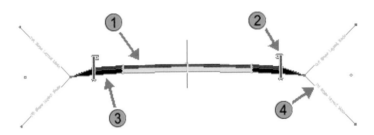

① Lane

② Guardrail

③ Shoulder

④ Daylight

Objectives

After completing this lesson, you will be able to:

- Describe how daylighting is used for matching slopes to surfaces.

- Describe subassembly components.

- Describe the process for modifying assembly properties.

- List the guidelines for creating and modifying transportation assemblies.

- Create an assembly that consists of lanes, shoulders, guardrails, and match slopes.

About Daylighting

As with assemblies used for subdivision corridor models, assemblies for transportation corridor models are made from subassemblies. The primary difference is that transportation corridor models usually involve daylighting to a surface, and the optional allowance for cut and fill ditches.

Definition of Daylighting

Daylighting is a function performed by a specific type of subassembly, where a slope is extended from that subassembly until it intersects with a surface. The point where this slope intersects the surface is called the daylight point, and if all these points were joined together as you traverse down the length of the alignment, the resultant line would be called the daylight line. This daylight line typically represents where the existing model and proposed model meet and is usually the limit of construction.

The grade (angle) that the slope line is extended upwards or downwards to intersect the existing surface is determined by the slope assembly's cut slope and fill slope parameter values. To extend the subassembly to the existing surface, the grade is determined by the cut slope and fill slope parameter values of the slope assembly. If the slope subassembly object is located above the existing surface at any particular station along the alignment, then it is considered to be in a fill condition and will project the slope downward until it intersects the existing surface using the value defined for its fill slope parameter. Conversely, if the slope assembly is located below the existing surface at any particular station, it will project a slope upward toward the existing surface based on its cut slope parameter value.

Some slope subassemblies can be very complex in their functionality. In general, these slope assemblies are designed to check for cut or fill conditions before projecting to calculate the daylight point. For example, a slope subassembly might have different cut slope values that depend on the depth of cut at that station. This subassembly might have maximum cut heights for flat, medium, and steep slopes built in. When Civil 3D® is evaluating the assembly at a particular station along the alignment, it iterates through these three cut height criteria to determine the appropriate slope to extend toward the existing surface. The subassembly first attempts to project a slope specified in the flat cut slope parameter. If that slope intersects the surface, and the height of the daylight point above the subassembly's hinge point is less than the flat cut maximum height, then the daylight point is created. If the height exceeds the allowable flat cut value, Civil 3D starts the process again and uses a steeper slope value as defined by the medium cut slope and projects to the surface. If it can find the surface within the height specified by the medium cut/maximum height parameter, the daylight point is established. If not, Civil 3D uses the steep cut slope value to project toward the surface and define the daylight line. This process is repeated at every station along the alignment where this subassembly exists.

How Daylighting is Calculated

The following illustration shows how a daylighting subassembly calculates the daylighting in both cut and fill conditions. In cut conditions, you can use different subassembly input parameters to vary the ditch configuration. You can also vary the cut slope based on the depth of cut. In fill conditions you can vary the fill slope based on height of fill.

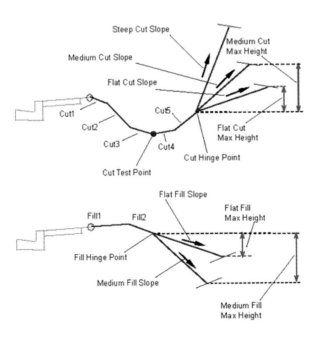

Chapter 12 | Transportation - Assemblies and Corridors

Example

The following illustration shows some daylighting subassemblies:

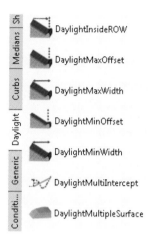

Subassembly Components

Subassemblies are made up of the following components: points, links, and shapes. Each component has unique code assignments that can be referenced for other purposes such as corridor display, design section labeling, and construction staking point generation.

There are many subassemblies that you use primarily to create transportation assemblies. In the example shown in the following illustration the user created a Transportation palette and copied subassemblies from the other palettes to it.

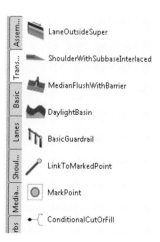

Code Set Style
The Code Set Style is a collection of marker styles, link styles, and shape styles. You can apply a code set style to assemblies, corridors, and design sections.

Subassembly Points

You use subassembly points when you:

- Generate corridor feature lines.

- Label offsets and elevations on design cross sections.

- Generate Civil 3D point objects that can be exported for construction staking.

Each subassembly point has a unique identifier. A marker style controls the display of subassembly points.

The following illustration shows subassembly points on the LaneOutsideSuper and BasicGuardrail subassemblies:

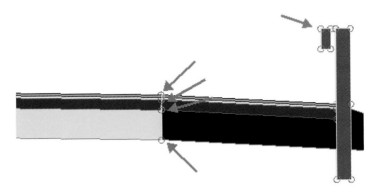

Subassembly Links

Subassembly links connect subassembly points. You use subassembly links when you:

- Create corridor surfaces.

- Label slopes on design cross sections.

Each subassembly link has a unique identifier. A link style controls the display of subassembly links. Subassembly links and are shown in the following illustration:

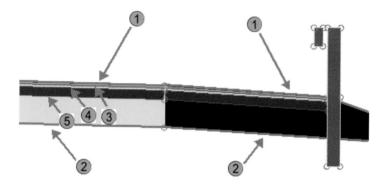

1. Top Pave Links

2. Datum and Subbase Links

3. Pave1

4. Pave2

5. Base

Subassembly Shapes

Subassembly shapes are closed areas defined by subassembly links. You use subassembly shapes when you:

- Calculate pavement structure volumes.

- Label pavement structure end areas on design cross sections.

Each subassembly shape has a unique identifier. A shape style controls the display of the subassembly shapes. Subassembly shapes are shown in the following illustration:

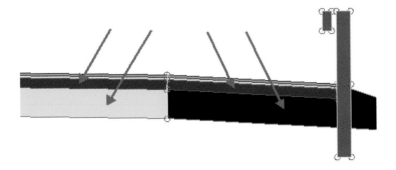

Subassembly Groups

Subassemblies on an assembly are organized in assembly groups. When you add subassemblies to an assembly, you can either select the assembly object, or you can select a marker point on a subassembly that was already added to the assembly. These are shown in the following illustration:

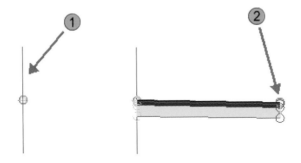

① Assembly (No Subassemblies)

② Subassembly Marker

When you select the assembly object to add the subassemblies, a new assembly group is created. When you select a subassembly marker to add a subassembly, the subassembly is added to an existing assembly group. Most subassemblies have a group for the left side subassemblies and the right side subassemblies. This is shown in the following illustration:

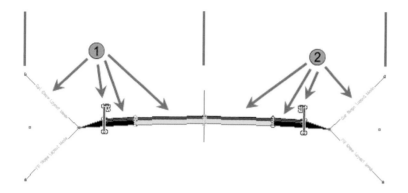

① Left Side Assembly Group

② Right Side Assembly Group

Modifying Assembly Properties

This section describes how you modify assembly properties.

Modifying Assembly Properties

When you modify the properties of the assembly, you can change subassembly input parameters such as lane width, cross slope, pavement structure depth, and daylight slope. Each subassembly has different input parameters.

You can also rename the assembly groups and the subassemblies within the assembly to organize the content of the assembly. The following illustration shows the Assembly Properties window with renamed assembly groups and renamed subassemblies:

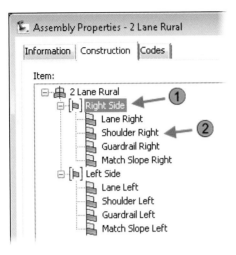

(1) Assembly Group

(2) Subassemblies

Finally, when you modify the properties of an assembly, you can change the appearance of the assembly by changing the code set style.

Guidelines for Creating Transportation Assemblies

Keep the following guidelines in mind when you create transportation assemblies.

- To simplify the assembly creation process, create a workspace that shows the Tool Palettes window on one side of the screen and the Properties window on the other side.

- Rename the subassemblies on the tool palette window to be more indicative of real life scenarios such as lane, shoulder, match, and so on.

- To mirror a subassembly from one side to another use the mirror command on the right click menu

- Modify the subassembly command settings so that when you add subassemblies to an assembly, the name of the subassembly references the local name, or the name on the tool palette (Lane, Shoulder, and Daylight), as opposed to the subassembly macro name (LaneOutsideSuper, ShoulderExtendedSubbase, and BasicSideSlopeCutDitch).

Example

The following illustration shows an assembly creation workspace that displays just the Properties and Tool Palettes windows. This workspace makes it easier and quicker to create an assembly.

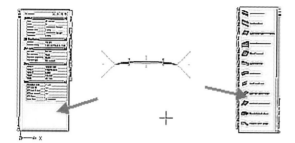

Exercise 01 | Create and Modify a Transportation Assembly

In this exercise, you create an assembly that consists of lanes, shoulders, guardrails, and match slopes. You then modify the properties of the assembly to rename the assembly groups and the subassemblies.

The completed exercise

Create Assembly

1 Open *Transportation - Assemblies and Corridors\I_create_assembly.dwg* (*M_create_assembly.dwg*).

2 To prepare your workspace:

- Close the Toolspace window.

- Open the Properties window by pressing CTRL + 1 on the keyboard.

- Dock the Properties window on the left side of the screen.

- Open the Tool Palettes window by pressing CTRL + 3 on the keyboard.

- Dock the Tool Palettes window on the right side of the screen.

3 Right-click the Tool Palette title bar. Select Civil Imperial Subassemblies (Civil Metric Subassemblies).

The imperial (metric) tool palettes is loaded.

✓	Civil Imperial Subassemblies
	Civil Metric Subassemblies
	Civil Materials
	Civil Multiview Blocks

4   On the ribbon, Home tab, Create Design panel, click Assembly > Create Assembly.

5   In the Create Assembly dialog box, For Name, enter **2 Lane Rural**. Click OK.

6   When prompted to Specify Assembly Baseline Location, select a location anywhere in the drawing area. Click to create the assembly.

The assembly object is created in the drawing.

Next, you attach the subassemblies to the assembly object.

7   On the Tool Palettes window:

- Click the Lanes tool palette.

- Click LaneOutsideSuper.

The Properties window is populated with the subassembly input parameters.

8   On the Properties window:

- Collapse the Information, General, and Data sections of the window.

- Under Advanced, for Side, select Right.

- For Width, enter **14' (3.75** m). Press ENTER.

- When prompted to Select Marker Point within Assembly, click the vertical line on the assembly to insert the right side lane.

9   On the Properties window, for Side, select Left.

10 When prompted to Select Marker Point within Assembly, click the vertical line on the assembly to insert the left side lane.

11 On the Tool Palettes window:

    • Click the Shoulders tool palette.

    • Click ShoulderExtendSubbase.

12 On the Properties window:

    • For side, select Right.

    • For Shoulder Width, enter **6' (2 m)**.

    • For Daylight slope, enter **3**.

13 When prompted to Select Marker Point within Assembly, zoom in and click the top outer right lane marker point to attach the right shoulder subassembly.

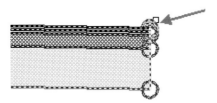

14 On the Object Properties window, for Side, select Left.

15   When prompted to Select Marker Point within Assembly, zoom to and click the top outer left lane marker point to attach the left shoulder subassembly. The assembly now contains the shoulders.

Note: You can use the Erase command to erase a subassembly if it was inserted on the wrong side.

Next, you add the guardrails.

16   On the Tool Palettes window:

- Click Basic tool palette.

- Click BasicGuardrail (Metric: BasicBarrier).

17   On the Properties window, for Side, select Right.

18   When prompted to Select Marker Point within Assembly, click the top marker on the outside edge on the right side of the shoulder pavement structure.

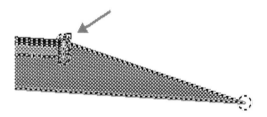

19   On the Properties window, for Side, select Left.

20   When prompted to Select Marker Point within Assembly, click the top marker on the outside edge of the shoulder pavement structure on the left side.

The guardrails (Metric: barriers) are now on both sides.

Next, you add the match slopes.

21   On the Basic tool palette, click BasicSideSlopeCutDitch.

22   On the Properties windows:

- For Side, select Right.

- For Cut slope, enter **2**.

- For Fill slope, enter **2**.

23   When prompted to Select Marker Point within Assembly, zoom in and click the subassembly marker on right side of the right shoulder.

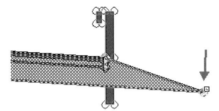

24   In the Properties window, for Side, select Left.

25   When prompted to Select Marker Point within Assembly, zoom in and click the outer subassembly marker on left side of the left shoulder.

26   Press ESC when finished.

The construction of the 2 Lane Rural assembly is complete.

27   Close the drawing. Do not save changes.

Modify Assembly

1   Open \Transportation - Assemblies and Corridors\I_assembly_properties.dwg (M_assembly_ properties.dwg).

2 Close the Properties and Tool Palettes windows.

3 On the ribbon, Home tab, Palettes panel, click Toolspace. Position the Toolspace window on the left side of the screen.

4 In Prospector:

- Expand Assemblies.

- Right-click 2 Lane Rural. Click Zoom To.

- Right click 2 Lane Rural. Click Properties.

5 In the Assembly Properties - 2 Lane Rural, Construction tab:

- Right-click Group (1) and rename as **Right Side**.

- Right-click Group (2) and rename as **Left Side**.

- Using the same procedure, rename the subassemblies as shown in the following illustration. (For metric users rename BasicBarrier to **Barrier**.)

- Click OK.

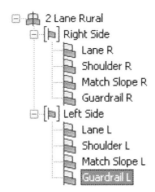

Subassemblies are organized into groups. A new group is created when you select the assembly object to add the subassemblies.

6 Close the drawing. Do not save changes.

Lesson 42 | Creating Transportation Corridors

This lesson describes how you create a corridor model for an arterial transportation facility. Transportation corridor models are usually more complicated than subdivision corridor models because they often include more complex subassemblies, daylighting to a surface, modeling lanes with varying widths and including superelevation.

A corridor model for an arterial roadway is shown in the following illustration:

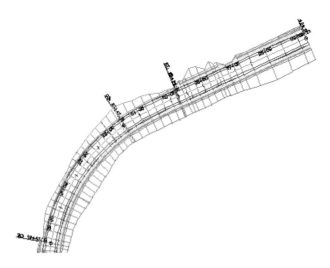

Objectives

After completing this lesson, you will be able to:

- Describe transportation corridors.

- Describe the process for mapping corridor targets.

- Explain how you create transportation corridor models.

- Create a corridor model.

- Map corridor targets.

- View and edit corridor sections.

About Transportation Corridors

Corridors used to model transportation facilities, such as intersections, interchanges, and road widenings, are usually more complex than corridors used to model subdivision roads.

Definition of Transportation Corridors

Transportation corridors are complex corridors that:

- Involve surface daylighting.

- Often read superelevation parameters assigned to an alignment.

- Require the use of other horizontal and vertical alignments or polyline geometry to control their configuration.

Subdivision corridor models use assemblies that reference basic lane, curb, and sidewalk subassemblies. In contrast, transportation corridor models use assemblies that reference a number of different subassemblies used for daylighting, shoulder, road reconstruction, and retaining walls.

Example

An example of a transportation corridor is shown in the following illustration.

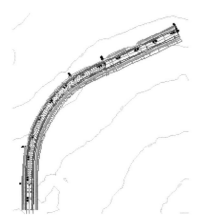

Target Mapping

When you create a corridor model, you assign a specific surface as a target parameter for the daylighting subassemblies. This process is called target mapping and is shown in the following illustration:

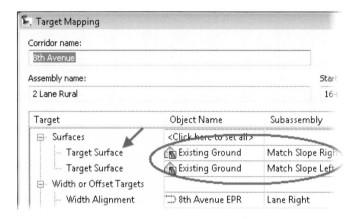

When you assign a surface as a target for the daylighting subassemblies, a feature line where the match slopes meet the surface is generated. This is called a *daylight feature line* and is shown in the following illustration:

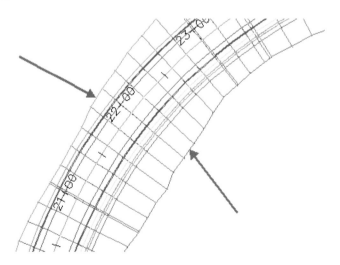

Target mapping for daylighting subassemblies is required because daylighting subassemblies does not work without assigning a surface. If you modify the subassembly properties and change the match slopes, or if the surface changes, the location of the daylight feature lines automatically recalculates.

Target Mapping Lane Widths

Many of the lane subassemblies can optionally use alignments and profiles as targets to override the default lane width and slope. The Help file for each subassembly indicates the target parameters. The following illustration shows the target segment of the Help file for the LaneOutsideSuper subassembly:

Parameter	Type	Description	Status
Width	Alignment	May be used to override the fixed lane Width and tie the edge-of-lane to an offset alignment.	Optional
Outside Elevation	Profile name	May be used to override the normal lane slope and tie the outer edge of the travel lane to the elevation of a profile.	Optional

You can model parallel lanes and tapers at intersection locations by assigning a target alignment at the edge of pavement location to the lane subassemblies in the assembly. The result is a corridor model that stretches the lane to accommodate the varying width through the taper and the fixed width at the parallel lane locations. This is shown in the following illustration:

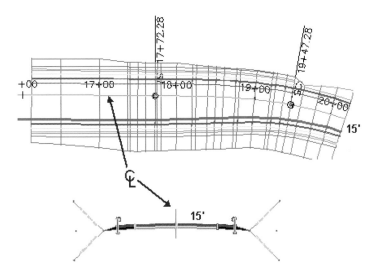

To assign corridor targets, you modify the corridor properties. The illustration shows the Lane Right subassembly targeting the 8th Avenue EPR alignment to vary the width. The illustration also shows the Match Slope subassemblies targeting the Existing Ground surface.

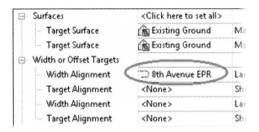

Creating Transportation Corridor Models

When you create a corridor model, you specify an alignment, a profile, and an assembly. These are the absolute minimum requirements for creating a corridor model.

Creating Transportation Corridors

A corridor region is the application of an assembly, or a typical cross section, over a station range. A corridor can have many corridor regions.

To create a corridor model, you can use the Create Simple Corridor command, or you can use the Create Corridor command. Each command results in the same corridor object. However, the Create Corridor command is generally used when you create corridors that model more complex cross-section configurations.

Most assemblies created for transportation corridor models use daylighting subassemblies to project match slopes to a surface in cut and fill conditions. The daylighting subassemblies use required target parameters to indicate which surface to project to. You can assign targets when you create the corridor, or when you modify the corridor properties.

Guidelines

Keep the following guidelines in mind when you create corridor models.

- When you create a corridor model, be aware of the station limits of the layout profile. For example, if an alignment is 1000 feet long, and the layout profile is 900 feet long, the corridor model should be calculated only over the length of the layout profile. When you modify the properties of a corridor model, change the starting and ending stations for the corridor model to coincide with the limits of the layout profile. When you do this, you do not receive corridor processing errors.

- When creating design profiles for corridor models, it is good practice to extend the design profiles beyond the starting and ending stations of the corridor model in order to eliminate corridor processing errors.

Exercise 01 | Create a Corridor Model

In this exercise, you create a corridor for an arterial road.

The completed exercise

1   Open \Transportation - Assemblies and Corridors\I_create_transportation_corridor.dwg
    (M_create_transportation_corridor.dwg).

2   On the ribbon, Home tab, Create Design panel, click Corridor > Create Corridor.

3   When prompted to select a baseline alignment, in the drawing area, click the 8th
    Avenue alignment.

4   When prompted to select a profile, press ENTER.

5   In the Select a Profile dialog box, for Select a Profile, select Design. Click OK.

6   When prompted to select an assembly, press ENTER.

7   In the Select and Assembly dialog box, confirm the assembly is 2 Lane Rural. Click OK.

8   In the Create Corridor dialog box, for Corridor Name, enter **8th Avenue**. Click Set All Targets.

9   In the Target Mapping dialog box:

   • In the Surfaces row, in the Object Name column, click in the value cell to set all.

   • In the Pick a Surface dialog box, select Existing Ground.

   • Click OK twice to close the dialog boxes.

10   Click OK to close the Create Corridor dialog box.

     The corridor object is created.

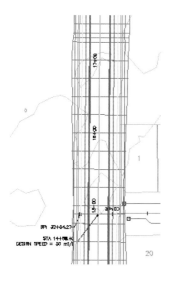

     Next, you analyze the corridor using the Object Viewer.

11   In the drawing area, select the corridor. Right- click, click Object Viewer.

12   In the Object Viewer window, change the view to SW Isometric.

13   In the Object Viewer window, zoom in on the corridor.

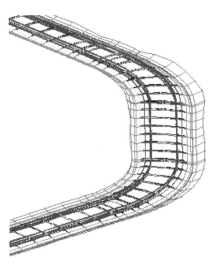

14   Experiment with the other settings in Object Viewer to view the corridor in 3D.

      Note: Do not click Set View. This button sets the drawing to the current view.

15   Close the Object Viewer window when you are done.

      Next, you split the corridor region.

16   In the drawing area, select the corridor. Right-click, click Corridor Properties.

17   In the Corridor Properties - 8th Avenue dialog box, Parameters tab, right-click region RG – 2 Lane Rural – (1). Click Split Region.

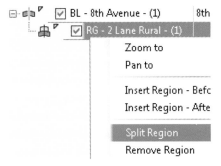

18   When prompted to Specify Station along Alignment, snap to the endpoint of the northwest corner of parcel number 1.

19   Press ENTER. Click OK. The corridor is split into two regions.

20   In the Corridor Properties dialog box, clear the region RG – 2 Lane Rural – (1) check box, to suppress the calculation of this region.

      Notice that clicking a region in the Corridor Properties dialog box highlights the region in the drawing area.

21   In the Corridor Properties dialog box, click Set All Frequencies.

22   In the Frequency to Apply Assemblies dialog box, under Apply Assembly:

      • For Along Tangents, enter **50' (20 m)**.

      • For Along curves, enter **10' (5 m)**.

      • For Along spirals, enter **10' (5 m)**.

      • For Along profile curves, enter **50' (20 m)**.

      • Click OK twice.

      The corridor to the north of the subdivision is modeled. The completed exercise drawing appears as follows:

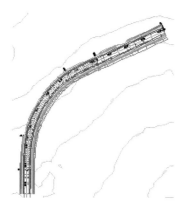

23   Close the drawing. Do not save the changes.

Exercise 02 | Map Corridor Targets

In this exercise, you model an acceleration lane and taper by mapping corridor targets.

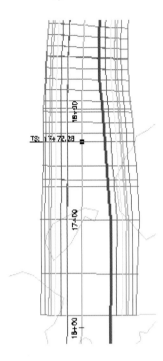

The completed exercise

1   Open *Transportation - Assemblies and Corridors\I_target_mapping.dwg*
    (*M_target_mapping.dwg*).

2   In the drawing area, note the 8th Avenue - EPR alignment.

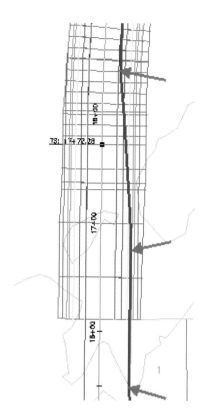

Next, you vary the width of the corridor lane with target mapping to the 8th Avenue - EPR alignment.

3   In the drawing area, select the corridor model.

4   On the contextual ribbon, Modify panel, click Corridor Properties.

5   In the Corridor Properties - 8th Avenue dialog box, Parameters tab, for Region (2), in the Target column, click the ellipsis.

6   In the Target Mapping dialog box, under Width or Offset Targets, for Width Alignment, click in the Object Name cell for the Lane Right subassembly.

7   In the Set Width Or Offset Target dialog box:

- For Select Object Type to Target, select Alignments.

- Click 8th Avenue EPR.

- Click Add.

8   Click OK three times to close the dialog boxes.

The corridor model rebuilds. It now displays the parallel lane and taper.

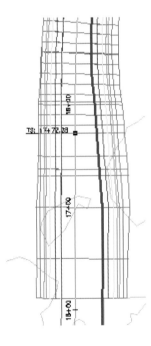

9   Close the drawing. Do not save the changes.

Exercise 03 | View and Edit Corridor Sections

In this exercise, you view and edit corridor sections.

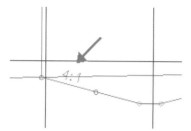

The completed exercise

1 Open \Transportation - Assemblies and Corridors\I_corridor_section_editor.dwg (M_corridor_section_editor.dwg).

2 Zoom in to the beginning of the corridor model.

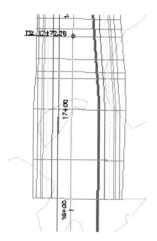

3 In the drawing area, select the 8th Avenue corridor model on the first corridor link at the beginning of the visible corridor.

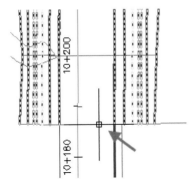

4   On the contextual ribbon click Modify panel > Corridor Section Editor.

The cross section is displayed at this location. The ribbon displays the Section Editor tools for the 8th Avenue corridor.

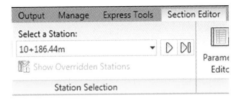

5   On the contextual ribbon, Station Selection panel:

• Click Next Station.

• View the corridor sections at the different station locations.

• Click Previous Station.

• Under Select a Station, use the station selector to view specific stations on the corridor.

6   On the ribbon, click Close to exit the Section Editor.

7   On the command line, enter **VPORTS**. Press ENTER.

8 In the Viewports dialog box, click Two: Horizontal. Click OK.

The screen divides into two model space viewports.

9 In the upper viewport, zoom to the extents of the corridor model.

10 In the lower viewport, select the 8th Avenue corridor.

11 On the contextual ribbon, click Modify panel > Corridor Section Editor.

The ribbon displays the Section Editor. In the lower viewport, the corridor section is displayed. In the upper viewport, a line is drawn across the corridor that represents the location of the current section.

12 On the contextual ribbon, Station Selection panel, click Next Station several times.

Note that the position of the line moves to represent the location of the current station.

13 Click Next Station until you view a section in the curve.

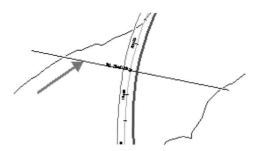

14 Note the superelevation in the cross section.

15 On the View/Edit Corridor Sections toolbar, change the station to **17+00' (10+220m)**.

Notice the widened lane on the right side.

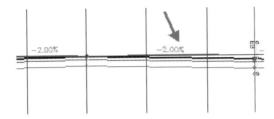

16 On the contextual ribbon click Corridor Edit Tools panel › Parameter Editor.

The Corridor Parameters editor is displayed in a separate window.

17 In the Corridor Parameters editor, under LEFT, Match Slope Left:

- For Cut Slope, for Value, enter **4**.

- For Fill Slope, for Value, enter **4**.

Notice that an override is applied to Cut Slope and Fill Slope.

	Cut Slope	4.00:1	ⓘ ☑ True	4.00:1
	Fill Slope	4.00:1	ⓘ ☑ True	4.00:1

Hover over the "i" symbol to see the default values.

18 Repeat the previous step and change the right cut and fill match slopes to **4:1**.

Next, you apply this override to a range of stations.

19 Move the Corridor Parameters window out of the way.

20 On the contextual ribbon, click Corridor Edit Tools panel > Apply to a Station Range.

21 In the Apply to a Range of Stations dialog box, for End station, enter **2300' (10440** m**)**. Click OK.

22 Select and view different stations. Notice where the slopes change from 2:1 to 4:1, and then back to 2:1 again.

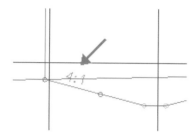

23 On the ribbon, click Close to exit the Section Editor.

24 In the drawing area, review the changes to the corridor model where the left and right side daylight slope transitions from 2:1 to 4:1, and then back to 2:1 again.

25 On the command line, enter **VPORTS**. Press ENTER.

26 In the Viewports dialog box, Click Single. Click OK.

27 Close the drawing. Do not save the changes.

Lesson 43 | Creating Transportation Corridor Surfaces

This lesson describes how you create transportation corridor surfaces.

Corridor surfaces can be used for earth cut and fill volume calculations, labeling finished design grades and slopes, and calculating pipe network rim and invert elevations.

A corridor surface is shown in 2D and 3D views in the following illustration.

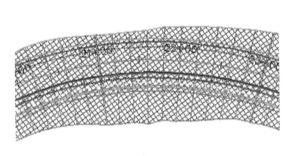
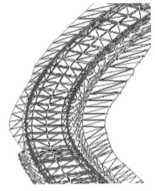

Objectives

After completing this lesson, you will be able to:

- Describe transportation corridor surfaces.

- Describe the process for creating a transportation corridor surface.

- Create corridor top and datum surfaces.

About Transportation Corridor Surfaces

You create corridor surfaces to represent finished grade and subgrade surfaces. Corridor surfaces that represent the subgrade are used for calculating earth cut and fill volume calculations.

Definition of Transportation Corridor Surfaces

Transportation corridor surfaces are created from the corridor links or the feature lines, and automatically update when the corridor model changes.

Uses of Transportation Corridor Surfaces

Corridor surfaces that represent the finished grade are used for the following:

- Labeling finished design spot elevations, grades, and slopes.

- Calculating rim and invert elevations for pipe networks.

- Modeling and creating surfaces that consist of both existing ground and finished design data.

Corridor surfaces that represent the subgrade are used for calculating earth cut and fill volume calculations.

Creating a Transportation Corridor Surface

This section outlines the steps for creating a transportation corridor surface.

Creating a Corridor Surface

When you create a corridor surface, you use either link data or feature line data. To create finished ground and subgrade corridor surfaces, you use the subassembly top links or the subassembly datum links.

Top and datum links are shown in the following illustration.

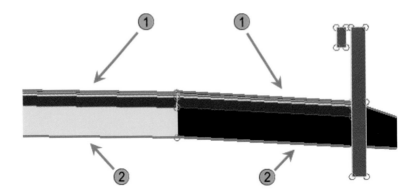

① Top Pave Links

② Datum and Subbase Links

Corridor Boundaries

When you create a corridor surface that represents the finished ground and the subgrade you must use boundaries to contain the triangulation within certain limits. Boundaries for finished ground and subgrade corridor surfaces are generally created from the corridor extents.

The following illustration shows a corridor surface without a boundary, created from the datum links:

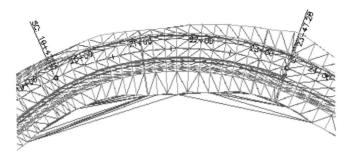

Boundary Options

To add corridor surface boundaries, you modify the properties of the corridor model. You can use the corridor extents to define the surface boundary, or you can create your own surface boundary.

Exercise 01 | Create Corridor Surfaces

In this exercise, you create corridor top and datum surfaces.

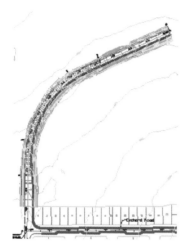

The completed exercise

1   Open *Transportation - Assemblies and Corridors\I_transportation_corridor_surface.dwg* (*M_transportation_corridor_surface.dwg*).

    First, you create a corridor datum surface.

2   In Prospector, expand Corridors. Right-click 8th Avenue. Click Properties.

3   In the Corridor Properties - 8th Avenue dialog box, Surfaces tab:

    • Click Create a Corridor Surface.

    • Rename the new surface **DATUM**.

    • For Data Type, select Links.

    • For Specify Code, select Datum.

    • Click Add Surface Item.

- For Datum Surface, click in the Surface Style cell.

- Select _Grid from the list.

- Click OK twice to close all dialog boxes.

The corridor datum surface is created and displayed using the _Grid surface style. Notice that the surface extends beyond the limit of the daylight lines.

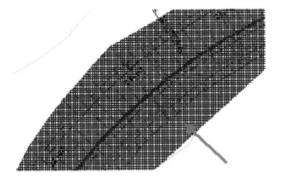

An outer boundary is required to limit the surface extents to the daylight lines. Next you define a boundary.

4   In Prospector, under Corridors, right-click 8th Avenue. Click Properties.

5   In the Corridor Properties - 8th Avenue dialog box, Boundaries tab:

- Right-click Datum. Click Add Corridor Extents.

- Click OK.

Note: If the option for Add Corridor Extents is not visible, click Add Automatically > Daylight, to select the daylight lines as the surface boundary.

The Daylight feature lines are used as an outside boundary for the datum surface. The surface is displayed correctly.

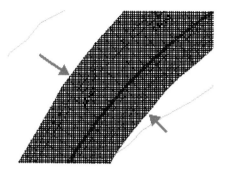

Next, you create a corridor surface that represents the top or finished surface. You follow the same procedures used to create the corridor datum surface.

6   In Prospector, under Corridors, right-click 8th Avenue. Click Properties.

7   In the Corridor Properties - 8th Avenue dialog box, Surfaces tab:

- For DATUM, under Surface Style, select _No Display.

- Click Create a Corridor Surface.

- Rename the new surface **TOP**.

- For Data Type, select Links.

- For Specify Code, select Top.

- Click Add Surface Item.

- For Top surface, Surface Style, click _No Display.

Next, you add outer boundary for the Top surface.

8   On the Boundaries tab:

- Right-click Top. Click Add Corridor Extents.
  Note: If the option for Add Corridor Extents is not visible, click Add
  Automatically > Daylight, to select the daylight lines as the surface boundary.

- Click OK.

Civil 3D creates the Top Corridor surface. The Top and Datum surfaces are not shown because
they were assigned the _No Display surface style.

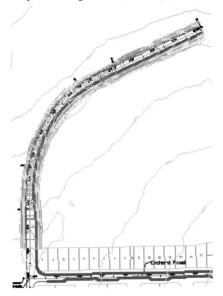

9   In Prospector, expand Surfaces.

Note the two corridor surfaces.

10   Close the drawing. Do not save the changes.

Lesson 44 | Creating Intersections

This lesson describes how to create a four-way intersection using the Create Intersection wizard. Intersection modeling can be very complex. Horizontal alignments, profiles, and assembly cross falls all require spatial coordination to correctly model an intersection. The Create Intersection wizard automatically generates and coordinates the alignments, profiles, corridor regions, and assemblies required to model the intersection and the entrances and exits. The end result is the creation of a corridor model and intersection object that are directly related to one another.

The following illustration shows an intersection with curb returns, viewed in Object Viewer.

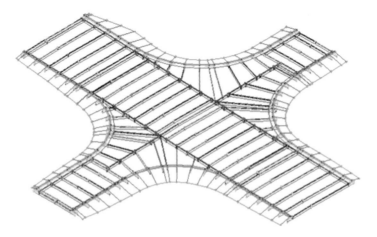

Objectives

After completing this lesson, you will be able to:

- Describe intersections.

- Explain the settings in the Create Intersection wizard.

- List the steps and guidelines for creating intersections.

- Create an intersection.

About Intersections

This section describes key aspects of intersection modeling in AutoCAD® Civil 3D®.

Definition of Intersections

Intersections represent the juncture of two roads. They are modeled with a corridor, and are the combination of horizontal and vertical geometry elements and corridor regions.

Driving Direction

When you model an intersection, you specify curb return parameters for the incoming and outgoing roads. A four way intersection is divided into four quadrants, each with an incoming and outgoing direction. The driving direction is used to identify the direction in which the quadrant curb returns are drawn when an intersection is created. The options are Right Side of the Road and Left Side of the Road. For left side driving, the curb return alignments in intersection objects are drawn starting on the left side of the outgoing road, and ending on the left side of the merging road.

To set the driving direction, you modify the Ambient Settings in Drawing Settings.

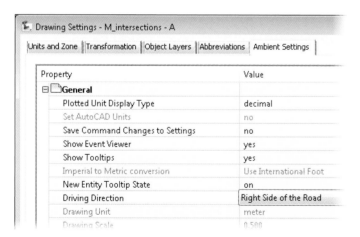

The default driving direction for a drawing is Right Side of the Road. In jurisdictions where people drive on the left side of the road, make sure you change this drawing setting.

Intersections Examples

The following illustrations show a three-way intersection and a four-way intersection created using the Create Intersection wizard.

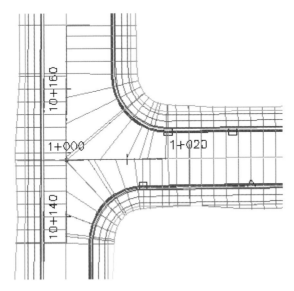

Three-way intersection

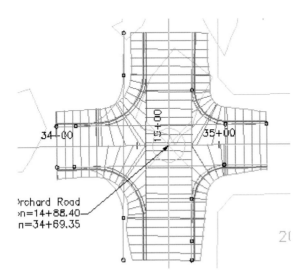

Four-way intersection

Create Intersection Wizard

In AutoCAD Civil 3D 2010, the task of creating an intersection has been simplified by the addition of the Create Intersection wizard. This section describes the three pages of the wizard you configure.

The Create Intersection wizard is launched from the ribbon.

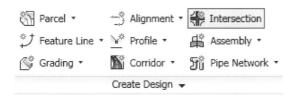

General Page

Use the General page to set the basic properties for the intersection, including the intersection name, an optional description, the intersection marker style, the intersection label style, and the intersection corridor type.

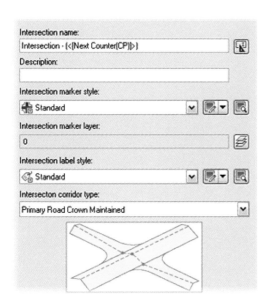

Corridor Type

The intersection corridor type value, of the Create Intersection wizard, determines the pavement elevations within the area of the intersection.

If the intersection corridor type is Primary Road Crown Maintained, the crown of the primary road is maintained, while the profile (crown) of the secondary road is adjusted to match the edge of the primary road and the intersection point. The crown (profile and edges) are not affected.

The following illustration shows the primary road crown maintained through the intersection.

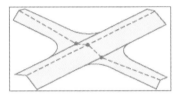

If the intersection corridor type is All Crowns Maintained, the profile of the side road is adjusted to match the main road elevation at the intersection point. The main road profile is not affected. Curb return profiles are generated to fit the profiles for the offset alignments.

The following illustration shows all crowns maintained through the intersection.

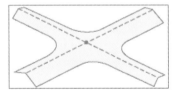

Geometry Details Page

Use the Geometry Details page to specify a variety of details about the geometry of the intersection object. You can:

- Change the priority of an alignment. Once the intersection is created, you cannot change the priority value.

- Specify the Offset Parameters to generate the left and right offset alignments for the primary and secondary roads.

- Specify Curb Return parameters for the geometry and radius of the curb return for each intersection quadrant.

- Specify Lane Slope parameters to calculate design profiles for the offset alignments.

- Specify Curb Return Profile parameters to calculate profiles for curb return alignments.

To set the geometry details for an intersection, you modify the Geometry Details page of the Create Intersection wizard.

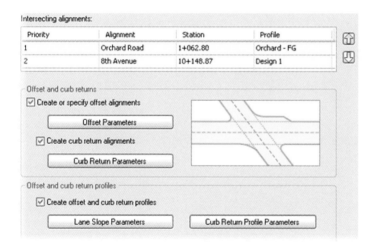

Corridor Regions Page

Use the Corridor Regions page to specify details about the corridor regions included in the intersection object. You can:

- Enable the Create Corridors in the Intersection Area option. When this option is selected, new corridor objects are created in the intersection area. When this option is not selected, the rest of the options on this page are not available for editing. You can create a new corridor. Or, you can add to an existing corridor, if your drawing contains corridors. You can also specify the surface to use for daylighting.

- Select an assembly set to import, and therefore use for creating the intersection.

- Specify the assembly that defines each section in a corridor region included in the intersection.

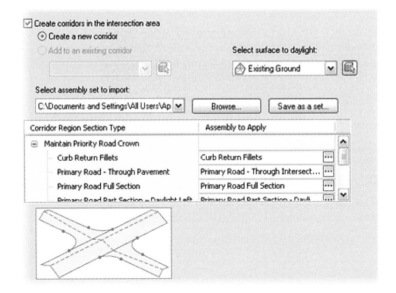

Assembly Sets

A corridor that models an intersection uses several assemblies to model the different areas of an intersection. Assembly sets can be loaded from an external file, or they can be saved in a drawing. You can also create your own custom assembly sets.

Chapter 12 | Transportation - Assemblies and Corridors

For example, an intersection corridor where the primary road crown is maintained makes use of an assembly set for the following intersection corridor region section types:

- Curb Return Fillets.

- Primary Road – Through Pavement.

- Primary Road Full Section.

- Primary Road Part Section – Daylight Left.

- Primary Road Part Section – Daylight Right.

- Secondary Road Full Section.

- Secondary Road Half Section – Daylight Left.

- Secondary Road Half Section – Daylight Right.

Creating Intersections

When you model intersections in AutoCAD Civil 3D, you create an intersection object and a related corridor model. First, you create the road geometry (centerlines and profiles) for the primary and secondary intersecting alignments. Second, you create an existing ground surface model targeted with the daylighting subassemblies used to model the intersection. Finally, you use the Create Intersection wizard to complete the intersection.

Process Description

To create an intersection, you can start with the following road geometry combinations:

- Two intersecting alignments and their design profiles.

- Two intersecting alignments and their design profiles with one or more road edge offset alignments or their profiles, or with both offset alignments and their profiles.

In the first scenario, the alignments and profiles for the offset and curb return alignments are automatically created with user input design parameters such as offset distances and radii. In the second scenario, the offset and curb return alignments and profiles are already created.

Process: Creating an Intersection

To create an intersection, you use the Create Intersection wizard.

1   Launch the Create Intersection command.

   • Select an intersection point for two alignments.

   • Select the main road alignment for which the road crown will be maintained.

   Note: When you create a four-way intersection, you must specify the primary (main) road alignment. If you are creating a three way (T-shaped) intersection, the through road is automatically selected as the primary road alignment.

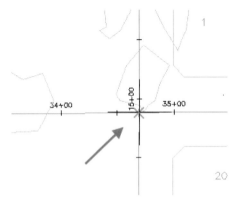

2   Configure the parameters in the General page of the Create Intersection wizard.

   • Provide an intersection name and description.

   • Select the intersection label styles.

   • Specify the corridor intersection type.

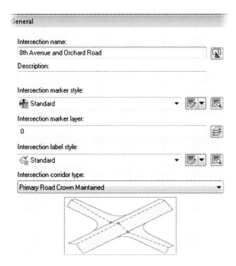

3 Configure the parameters in the Geometry Details page of the wizard.

- Adjust the alignment priority, if necessary.

- Specify the Offset Parameters to generate the left and right offset alignments for the primary and secondary roads.

- Specify Curb Return Parameters for the geometry and radius of the curb return for each intersection quadrant.

- Specify Lane Slope Parameters to calculate design profiles for the offset alignments.

- Specify Curb Return Profile parameters to calculate profiles for curb return alignments.

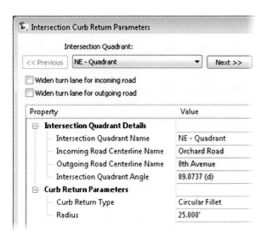

4   Configure the parameters in the Corridor Regions page of the wizard.

- If you are creating corridors for the intersection area, determine if you will create a new corridor or add to an existing one.

- Select a surface to daylight to.

- Select an assembly set to import.

- Confirm the assembly assigned to each corridor region section type.

Guidelines

Keep the following guidelines in mind when creating intersections.

- For a four-way intersection, the alignments must physically cross over each other so that an intersection point can be located.

- For a three-way intersection, the endpoint of the secondary road alignment must snap to a point on the primary road alignment. If the secondary alignment extends across the primary alignment, Civil 3D attempts to model a four-way intersection.

- The assembly set must be manually selected in the Create Intersection wizard to ensure the correct imperial or metric assemblies are used.

- When creating a corridor without the Create Intersection wizard, the intersection object and the offset curb return alignments and profiles are created. You can create the corridor regions as an independent step.

- If you are creating an intersection for jurisdictions where people drive on the left side of the road, make sure you change the Driving Direction in the Drawing Settings.

Exercise 01 | Create an Intersection

In this exercise, you create a four-way intersection using the Create Intersection wizard. You start with two intersecting alignments and profiles.

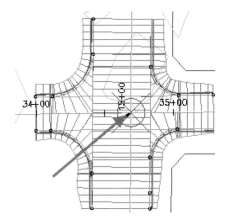

The completed exercise

1   Open *\Transportation - Assemblies and Corridors\I_intersections.dwg (M_intersections.dwg)*.

2   On the ribbon, click the Home tab > Create Design panel > Intersection.

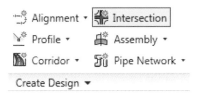

3   When prompted to Select Intersection Point, in the drawing area, click the intersection of the 8th Avenue and Orchard Road alignments.

    Note: The intersection OSNAP is automatically turned on.

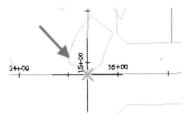

4 When prompted to Select Main Road Alignment, in the drawing area, click the 8th Avenue alignment. This is the alignment that runs from southwest to northeast.

The Create Intersection wizard begins.

5 On the General page:

- For Intersection Name, enter **8th Avenue and Orchard Road**.

- Verify that Intersection Corridor Type is set to Primary Road Crown Maintained.

- Click Next.

6 On the Geometry Details page, click Offset Parameters.

The Offset Parameters are used to generate the offset alignment pavement edges, measured from the road centerline.

7 In the Intersection Offset Parameters dialog box, under Primary Road, for:

- Left Offset Alignment Definition, change Offset Value to **14' (4 m)**.

- Right Offset Alignment Definitions, change Offset Value to **14' (4 m)**.

- Click OK.

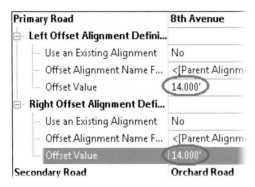

8   On the Geometry Details page, click Curb Return Parameters.

Curb return parameters are used to specify the radii of the intersection curb returns. In this example, the curb returns on the entrance to the subdivision on the east side are smaller than the curb return radii on the west side.

9   In the Intersection Curb Return Parameters dialog box:

- Verify that Intersection Quadrant is set to NE - Quadrant.

- Change Radius to **20' (10 m)**.

- Click Next.

- Change Intersection Quadrant to SE - Quadrant.

- Change Radius to **20' (10 m)**.

- Click Next.

- Change Intersection Quadrant to SW - Quadrant.

- Change Radius to **30' (12 m)**.

- Click Next.

- Change Intersection Quadrant to NW - Quadrant.

- Change Radius to **30' (12 m)**.

- Click OK.

Note: Notice the preview graphics in the drawing area as you change quadrants.

10   The Lane Slope Parameters are used to generate the profiles for the offset alignments, beyond the extents of the curb returns. On the Geometry Details page, click Lane Slope Parameters.

11   In the Intersection Lane Slope Parameters dialog box, review the settings. Click OK.

12   On the Geometry Details page, click Curb Return Profile Parameters.

The curb return profile parameters are used calculate the profiles for the curb returns. They are also used to extend the curb return profiles along the incoming and outgoing lanes beyond the extents of the curb returns.

13   In the Intersection Curb Return Profile Parameters dialog box, review the settings. Click OK.

14   On the Geometry Details page, click Next.

15   On the Corridor Regions page, click the Browse button.

16   In the Select Assembly Set File dialog box:

   • Click the up arrow.

   • Notice the Imperial and Metric folders.

   • Click Imperial (Metric).

   • Click _Autodesk (Imperial) Assembly Sets.xml (_Autodesk (Metric) Assembly Sets.xml).

   • Click Open.

17   Click Create Intersection.

18   Close Panorama if it appears.

     A corridor model showing an intersection is displayed in the drawing area.

19   Using grips, drag the intersection label away from the intersection area.

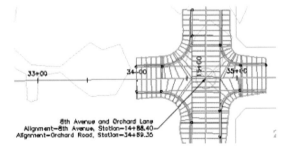

8th Avenue and Orchard Lane
Alignment–8th Avenue, Station–14+88.40
Alignment–Orchard Road, Station–34+59.36

     Next, you review the data.

20   In Prospector, expand Corridors. Right-click Corridor – (1). Click Properties.

21   In the Corridor Properties dialog box, Information tab, for Name, enter **INT 8th and Orchard**.

22   Click the Parameters tab. Note the construction of the corridor. It consists of multiple baselines and regions. Click OK.

23 In Prospector, expand Alignments and Offset Alignments.

Note the 4 offset alignments. Each offset alignment has a design profile.

24 In Prospector:

- Expand Alignments, Curb Return Alignments.

- Note the 4 curb return alignments. Each curb return alignment has a design profile.

25 In Prospector, expand all the trees under Intersections and review the data.

Note the references to the 8th Avenue and Orchard Road centerline and offset alignments. Also note the reference to the intersection curb return alignments for each quadrant. The Intersection collection shows, as references, all the alignments used to construct the intersection. In this collection you can modify the properties of the alignments and the parameters used to create the offset and curb return alignments.

26 In the drawing area, navigate to the Orchard Road (Secondary Alignment) Profile View.

Notice that the profile has been adjusted to match the 8th Avenue Profile.

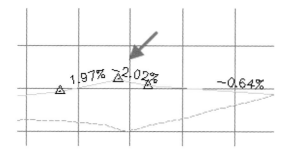

Note: The grades of the primary road cross fall are not exactly 2% because the intersecting alignments are slightly skewed.

Next, you now make a design change to the intersection.

27 In the drawing area, select the layout profile.

Notice the locks on the PVI's at the intersection with 8th Avenue.

28   In Prospector:

   - Expand Intersections, 8th Avenue and Orchard Road, Intersecting Alignments and 8th Avenue.

   - Right-click 8th Avenue-Left-14.000 (8th Avenue-Left-3.500).

   - Click Edit Offset Parameters.

29   In the Intersection Offset Parameters dialog box, for Primary Road, Left Offset Alignment Definition, change the Offset Value to **28' (7 m)**.

30   Close the Intersection Offset Parameters palette.

31   In the drawing area, the offset and curb return alignments shift to the left.

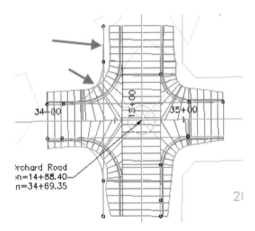

32   In Prospector, note that the intersection and corridor are out of date.

33   In Prospector, under Intersections, right-click 8th Avenue and Orchard Road. Click Recreate Corridor Regions.

34   In the Intersection Corridor Regions dialog box:

   - For Select Surface to Daylight, click Existing Ground.

   - Click Recreate.

35   Close Panorama if it is displayed.

   The corridor and intersection are no longer out of date and the drawing is updated to reflect the new offsets.

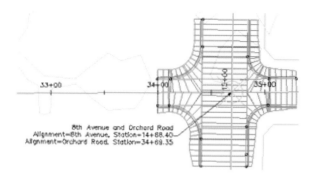

8th Avenue and Orchard Road
Alignment=8th Avenue, Station=14+88.40
Alignment=Orchard Road, Station=34+69.35

36   Close the drawing. Do not save the changes.

Lesson 45 | Modeling Road Designs in 3D

This lesson describes how you create a model that combines an existing surface and a corridor model. It also describes how you view and render a model with photorealistic materials in 3D.

Three-dimensional models of a proposed design are very effective when you need to communicate the plan or the design to both the general public and approving agencies. You can easily apply photorealistic materials to corridors that are rendered in 3D for presentation purposes. Furthermore, you can merge the corridor model with the existing ground surface model to enhance the effect.

The following illustration shows a 3D view of a road design.

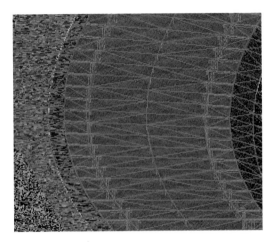

Objectives

After completing this lesson, you will be able to:

- Describe how a code set style assigns rendered material styles to corridor links.

- Describe the process for creating a 3D road design model.

- Create a 3D road design model.

About Code Set Styles

A corridor is rendered based on the rendered material style assigned to the corridor links. The code set style assigns rendered material styles to the corridor links.

Definition of Code Set Styles

A code set style is a collection of the following styles:

- Marker

- Link

- Shape

- Render material

- Label

- Feature line

A code set style is applied to an assembly, a corridor, and corridor sections. The code set style assigns the render material styles to corridor links. Each link can have a different render material assigned. This means that you do not need to create a surface to create photorealistic rendered images.

Code Set Style Example

The following illustration shows the code set styles for various corridor links:

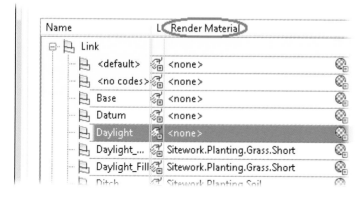

Creating a 3D Road Design Model

You create 3D models that combine proposed and existing conditions to help visualize and analyze the impacts your design has on the existing terrain. These models can also be used to create perspective views from different view points. The process to create a 3D corridor design model that shows both existing and proposed conditions is as follows:

- Create a new surface called EG FG Combined.

- Cut a hole in the surface using the daylight lines as a boundary.

- View and render the EG FG Combined surface and corridor model in 3D.

The surface is rendered based on the rendered material style assigned to the surface. The rendered material style is a property of the surface and is shown in the following illustration.

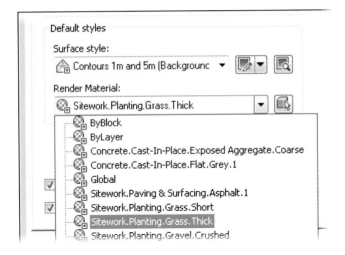

Creating a 3D Road Design Model

The first step in creating a 3D road design model is to create a new surface called *EG FG Combined*. After you create the EG FG Combined surface, you edit the surface and paste the existing ground surface into the EG FG Combined surface. Now that you have data attached to the surface, you can begin the process for cutting a hole in the surface to remove the surface data inside the corridor model daylight lines. The steps are as follows:

- Create 3D polylines from the daylight feature lines.

- Turn the 3D polylines into 2D polylines and form a closed polyline.

- Add the closed polyline as a hide boundary to the EG FG Combined surface (this deletes the surface data inside the closed polyline).

The final step is to view the corridor model and the EG FG Combined surface in 3D and apply the rendering. In Object Viewer, you change the view to Realistic to show a rendered image.

Exercise 01 | Create a 3D Road Design Model

In this exercise, you create a 3D model of the completed road design.

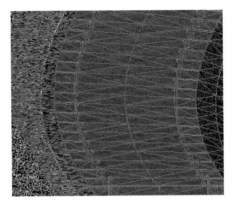

The completed exercise

1   Open *Transportation - Assemblies and Corridors\\I_3d_model.dwg (M_3d_model.dwg)*.

First, you make a copy of EG surface.

2   In Prospector, expand Surfaces. Right-click Existing Ground. Click Select.

3   On the command line, enter **CO** (Copy). Press ENTER.

4   Press ENTER twice to end the command.

5   In Prospector:

   • Expand Surfaces. The new surface is displayed as Existing Ground (1).

   • Right-click Existing Ground(1). Click Surface Properties.

6   In the Surface Properties Existing Ground (1) dialog box:

   • Clear the Object Locked check box.

   • For Name, enter **EG FG Combined**.

   • For Surface Style, select Contours 1' and 5' (Background) (Contours 1 m and 5 m (Background)).

   • Click OK.

Next, you suppress the display of the other surfaces.

7   In Prospector:

• Click Surfaces.

• In the item view area, for Existing Ground, for Style, select _No Display.

• In the item view area, for TOP, for Style, select _No Display.

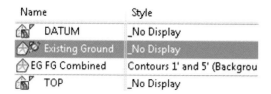

Note: You can change the style for multiple surfaces. Using CTRL, select the surfaces in the item view area. Right-click over the Style column header and click Edit. Select a surface style.

Next, you paste the TOP surface into the EG FG Combined surface.

8   In Prospector:

• Expand Surfaces, EG FG Combined, Definition.

• Right-click Edits. Click Paste Surface.

9   In the Select Surface to Paste dialog box, select Top. Click OK.

The 8th Avenue Corridor Top surface is pasted into the EG FG Combined surface.

10 Zoom to the 8th Avenue Corridor near the middle of the curve.

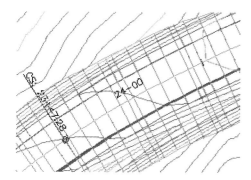

Notice how the contours reflect the design elevations for 8th Avenue inside the corridor daylight lines, and the existing surface elevations outside the corridor daylight lines.

11   In the drawing area:

  • Select the EG FG Combined surface.

  • Right-click. Click Object Viewer.

12   In the Object Viewer:

  • For Display Mode, select 3D Wireframe.

  • For View, select SW Isometric.

13   Use the tools in the Object Viewer to display the EG FG Combined surface in 3D.

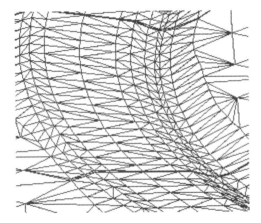

14   Close Object Viewer.

      Next, you render the corridor model.

15   In the drawing area:

  • Select the corridor.

  • Right-click, click Corridor Properties.

16   In the Corridor Properties - 8th Avenue dialog box, Codes tab, for Code Set Style, select Rural Corridor Rendered. Click OK.

17   In Prospector, right-click EG FG Combined. Select Rebuild - Automatic.

Next, you assign a render material to the EG FG Combined surface.

18 In the drawing area:

- Select the surface.

- Right-click, click Surface Properties.

19 In the Surface Properties EG FG Combined dialog box, Information tab, for Render Material, select Sitework.Planting.Grass.Short. Click OK.

20 In the drawing area:

- Select the 8th Avenue Corridor.

- Right-click, click Object Viewer.

21 In Object Viewer:

- For Display Mode, select Realistic.

- For View, select SW Isometric.

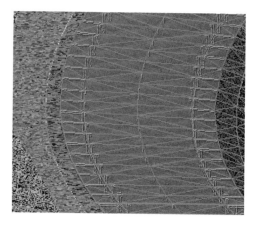

22 Experiment with other views in Object Viewer.

23 Close the Object Viewer. Close the drawing. Do not save the changes.

Chapter 13

Transportation - Sections and Quantities

In this chapter, you create design sections and calculate quantities for corridor models. The first step in the process is to create sample lines at intervals along the alignment. Sample lines are organized in a sample line group and are required for creating design cross sections and calculating quantities.

When you create sample lines, you attach section data to the sample lines. The section data you attach can then be viewed in the design cross sections or used to calculate quantities for the corridor model. You can attach surface, corridor surface, corridor, material, and pipe network section data to the sample lines.

After you create the sample lines, you calculate quantities for the corridor model. Earth cut and fill quantities are calculated by comparing the corridor datum surface with the existing ground surface. Corridor pavement structure quantities are calculated from the corridor section data. Quantities are reported either in tables in the drawing area or in a web browser. Quantity tables in the drawing automatically update when you make changes to the corridor model. This is useful when you want to adjust the layout profile to balance cut and fill or achieve a material excess or deficit.

Sample lines are also used to show design cross sections at the different station locations along the alignment. Section views are the grid objects that show section data. Section views are created and organized in the drawing using a group plot style. The group plot style orients the section views in arrays of rows, columns, or within sheets. When you create the section views, you indicate which section data to display.

▶ **Objectives**

After completing this chapter, you will be able to:

- Create sample lines for alignments.

- Calculate earthworks cut and fill and pavement structure quantities for a corridor model.

- Create quantity reports from sample line groups.

- Create sections to view elevation data across an alignment.

Lesson 46 | Creating Sample Lines

This lesson describes how you create sample lines.

Sample lines are required to calculate quantities and to create section views that display section data. Sample lines are attached to an alignment, as shown in the following illustration.

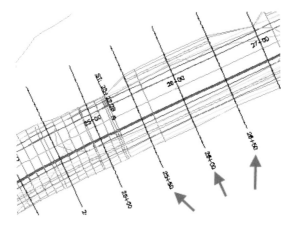

Objectives

After completing this lesson, you will be able to:

- Describe sample lines and how they are used in cross sections.

- Describe sample line parameters.

- Describe how you create sample lines from an alignment.

- Create and edit sample lines.

- Modify the sample line group properties and add additional section data.

About Sample Lines

Sample lines are created for an alignment and extract section data for surfaces, corridors, and pipe networks. They represent the locations where section data is generated.

Definition of Sample Lines

Sample lines are linear objects with a specific width that are used to cut sections at a specified interval along an alignment. Sample lines are created along the corridor baseline and are used to:

- Calculate earth cut/fill and pavement structure corridor quantities.

- Create section views.

Sample Line Groups

Sample lines are organized in sample line groups. Sample line groups contain the following elements:

- Sample line data.

- Section data.

- Section view groups.

- Mass haul lines and views.

You can perform the following tasks when you modify the properties of a sample line group:

- Rename the sample line group.

- Edit the swath width and change the labels assigned to the sample lines.

- Add or remove section data from the sample lines.

- Modify properties of the section views.

- Modify the material list.

The last two functions are applicable only after you create section views and calculate the quantities for the corridor.

Example

The following illustration shows a single sample line placed on an alignment. The name SL-1 was generated automatically.

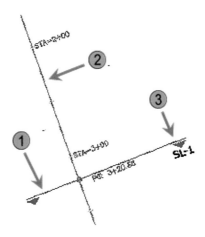

① Sample line

② Alignment

③ Sample line label

Sample Line Parameters

To create sample lines, you specify the data source and location parameters.

Data Source

The section data you attach to the sample lines is used to calculate quantities and can be displayed on section views. You can attach the following types of section data to sample lines:

Data source	Description
Surface Section Data	Existing ground or other surface data used for calculating earthworks volumes and displaying in section views.
Corridor Surface Section Data	A corridor datum surface that can be used for volume calculations and optionally displayed in section views.
Corridor Section Data	Use for calculating pavement structure quantities and for displaying in section views.
Pipe Network Section Data	Use for displaying pipe networks in section views.
Material Section Data	Created after volumes have been calculated and used to show material areas in section views.

Location

You create sample lines by specifying a range of stations and the width of the swath to the left and the right of the alignment. When you create sample lines, you can specify their location using the following methods.

Location	Description
From corridor stations	Sample lines created at the locations where the assembly was inserted to create the corridor.
By range of stations	User-specified sample line increment, independent of the corridor assembly insertion frequency.
At a station	User-specified individual stations, such as driveways, entrances, and BC and EC of intersection curb returns.
Pick points on screen	Create skewed sample lines or multivertex sample lines.

Creating Sample Lines

You create sample lines to:

- Create design cross sections at intervals along an alignment.

- Calculate corridor earth cut and fill and pavement structure quantities.

Creating Sample Lines

To create sample lines, you first select an alignment. You then specify the data that is to be attached to the sample lines. You select the section data to attach to the sample lines based on data required for quantity calculations, and data you want to view in the section view objects. The next step is to determine where the sample lines are created along the alignment. Finally, you specify the width of the sample lines. The sample line width can either be a fixed value or can be determined from another alignment.

Moving Sample Lines

You can edit the locations of sample lines in plan view by using grips. The sample line grips are shown in the following illustration.

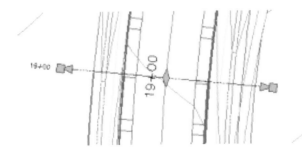

The grips are summarized as follows:

- Diamond grip repositions the sample line along the alignment.

- Triangle grips change the length (swath width) of the sample line.

- Square grips change the length and skew angle of the sample line.

When you change the location of a sample line, the data attached to the sample line is updated based on the new location. Quantities tables in the drawing and cross section views also automatically update.

Exercise 01 | Create and Edit Sample Lines

In this exercise, you create and edit sample lines.

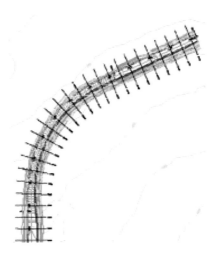

The completed exercise

1 Open...*Transportation - Sections and Quantities_sample_lines.dwg (M_sample_lines.dwg)*.

2 On the ribbon, click the Home tab > Profiles & Section Views panel > Sample Lines.

3 When prompted to Select an Alignment, press ENTER.

4 In the Select Alignment dialog box, click 8th Avenue. Click OK.

5 In the Create Sample Line Group dialog box:

- Clear the 8th Avenue TOP check box.

- For 8th Avenue DATUM, click in the Style column.

- In the Pick Section style dialog box, select Datum.

- Click OK twice.

6   On the Sample Line Tools toolbar, select By Range of Stations.

7   In the Create Sample Lines - By Station Range dialog box:

- Under Left Swath Width, for Width, enter **70' (25 m)**.

- Under Right Swath Width, for Width, enter **70' (25 m)**.

- Under Sampling Increments, for Increment Along Tangents, enter **50' (20 m)**.

- For Increment Along Curves, enter **50' (20 m)**.

- For Increment Along Spirals, enter **50' (20 m)**.

- Click OK.

8   Close the Sample Line Tools toolbar. The sample lines are created in the drawing area.

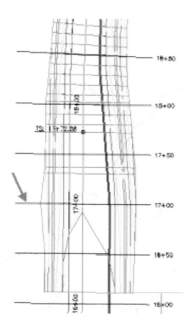

9   In Prospector:

- Expand Alignments, Centerline Alignments, 8th Avenue, Sample Line Groups,
  SL Collection - 1.

Notice that the sample line group can contain sample lines, sections, section view groups, mass haul lines, and mass haul views. Also notice that just the Sample Lines and Section collections are populated with data.

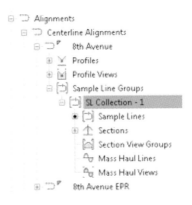

You can graphically modify the location of the sample lines.

10 In the drawing area:

- Select any sample line.

- Click the diamond shaped grip.

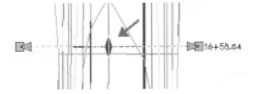

- Move your mouse along the alignment and click a new location.

The label on the sample line position changes and the sample line updates to reflect the new station location.

11 The completed exercise appears as follows:

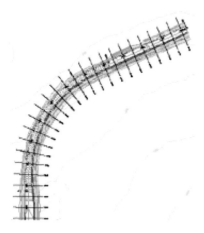

12 Close the drawing. Do not save the changes.

Exercise 02 | Modify Sample Line Group Properties

In this exercise, you modify the sample line group properties and add additional section data.

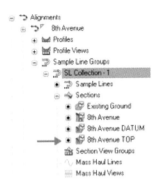

The completed exercise

1   Open \Transportation - Sections and Quantities\I_slg_properties.dwg (M_slg_properties.dwg).

2   In Prospector:

• Expand Alignments, Centerline Alignments, 8th Avenue, Sample Line Groups and SL Collection -1.

• Expand Sections.

• Note the Existing Ground surface, 8th Avenue corridor, and 8th Avenue DATUM corridor surface section data are attached to the sample lines. This data can be used for quantity calculation and displayed in the section views.

3   Right-click SL Collection - 1. Click Properties.

4   In the Sample Line Group Properties - SL Collection - 1 dialog box, click the Sample Lines tab. Note that you can change the offsets and the labels for the sample lines.

5   Click on the Sections tab and:

• Note the tools that enable you to add or remove section data to the sample lines. When you add section data to the sample lines, this data can be used for quantity calculations and displayed in the section views.

• Click Sample More Sources.

• On the left side of the Section Sources dialog box, click 8th Avenue TOP.

- Click Add to add the 8th Avenue TOP corridor surface to the sampled sources data list.

- For 8th Avenue Top, change the Style to Finished Ground.

- Click OK twice. The 8th Avenue TOP corridor surface is added to the sample lines with the Finished Ground style assigned.

6   Click the Section Views tab and then click the Material List tab. No data is displayed on these tabs. Section views have not been created and material quantities have not been computed.

7   Click OK to close the dialog box.

8   In Prospector, note that the Section data has been updated to include the 8th Avenue TOP corridor surface section data.

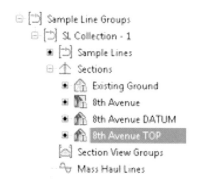

You can also access sample line group properties from the drawing area.

9   In the drawing area, select any sample line.

10   On the contextual ribbon, click the Modify panel > Group Properties.

11   In the Sample Line Group Properties - SL Collection - 1 dialog box, review the settings. Click OK to close the dialog box.

12   Close the drawing. Do not save the changes.

Lesson 47 | Calculating Corridor Quantities

This lesson describes how you calculate earth cut and fill and pavement structure quantities from sample lines. Earth cut and fill volumes are calculated by comparing the corridor datum surface section data with the existing ground surface section data. Pavement structure volumes are calculated directly from the corridor section data. The reporting of corridor quantities is a separate process.

You calculate corridor quantities to estimate the amount of earth to be moved and the required materials for construction. For example, you can use surfaces generated by a corridor model that represent roadway subgrades to estimate how much earth needs to be moved. You use the corridor section data to calculate the quantity of asphalt and granular material required to construct the roadway.

The following illustration shows a material list for corridor quantities. The material list is assigned to a sample line group.

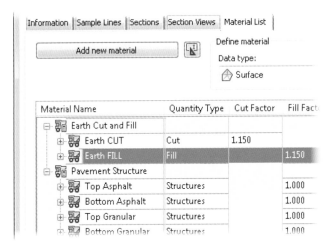

Objectives

After completing this lesson, you will be able to:

- Describe criteria used in quantity takeoff calculations.

- Explain how you calculate corridor quantity takeoff.

- Calculate the earth cut and fill and pavement structure quantities for a corridor model.

About Quantity Takeoff Criteria

You use quantity takeoff criteria to indicate exactly how the quantities are to be calculated. You can specify the name of the material, the data used to calculate the material quantity, the condition, and the quantity type. For earth cut and fill quantities, you can also specify cut factors, fill factors, and refill factors.

To calculate corridor quantities, you first create a quantity takeoff criteria.

Definition of Quantity Takeoff Criteria

Quantity takeoff criteria indicate precisely how the quantities are to be calculated. Quantity takeoff criteria are defined in the drawing template and can be modified to suit specific project requirements.

Criteria Properties

When you create quantity takeoff criteria, you specify the following information:

Item	Description
Material name	Name of the material to appear in the report.
Data type	Surface or pavement structure.
Condition	Above or below.
Quantity type	Cut, fill, cut and refill, earthworks, structures.
Cut factor	Earth cut material expansion factor.
Fill factor	Earth fill material compaction factor.
Refill factor	Percentage of earth cut material that is suitable for engineering fill material.
Shape style	Controls the display of the material in the cross section views.

Cut and Fill Factors

The cut factor is also referred to as an *expansion* factor. If you measure 100 cubic yards of excavation, then you actually excavate 115 (1.15 factor) cubic yards because the material expands after excavating.

Conversely, a fill factor is also known as a *compaction* factor. If you measure 100 cubic yards of fill, you actually require 115 (1.15 factor) cubic yards of material because of compaction. The refill factor is the percentage of cut material suitable for engineering fill material.

Example

The following illustration shows examples of quantity takeoff criteria used to calculate earth cut and fill quantities.

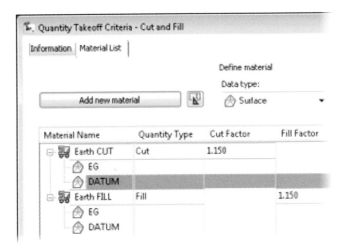

Calculating Corridor Quantities

To calculate corridor quantities, you first create a quantity takeoff criteria. After you create the quantity takeoff criteria, you then compute the material volumes. Earth cut and fill volumes are calculated using the existing ground and corridor surface section data attached to sample lines. Pavement structure volumes are calculated using the corridor section data attached to sample lines.

Calculating Corridor Quantities

A quantity takeoff criteria is created in your company/client drawing template, and can be used to calculate the quantities for most projects. You can also create new quantity takeoff criteria in the current drawing for projects that have unique quantity calculation requirements.

A quantity takeoff criteria for pavement structure is shown in the following illustration.

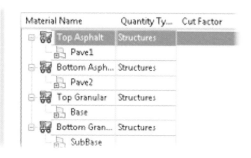

After you create or review the quantity takeoff criteria, you then compute the material quantities as a single step.

When you compute the material quantities, a material list is attached to the sample line group. This is shown in the following illustration. When the corridor model changes, the material list, and hence the volumes, automatically updates.

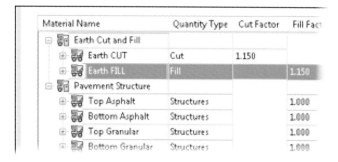

Guidelines

Keep the following guidelines in mind when calculating corridor quantities.

- A common practice is to create individual quantity takeoff criteria that can be used for earth cut/fill volumes and for pavement structure volumes. This simplifies and organizes the reporting process.

- When you compute materials, no quantities are displayed. To view the quantities you must generate a report.

Exercise 01 | Calculate Corridor Quantities

In this exercise, you review the quantity takeoff criteria, and calculate the earth cut and fill and pavement structure quantities for a corridor model.

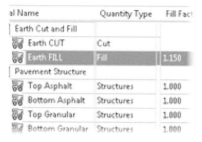

The completed exercise

1   Open *Transportation - Sections and Quantities\I_corridor_quantities.dwg* (*M_corridor_quantities.dwg*).

2   In Toolspace, Settings tab:

   • Expand Quantity Takeoff and Quantity Takeoff Criteria.

   • Right-click Cut and Fill. Click Edit.

3   In the Quantity Takeoff Criteria - Cut and Fill dialog box, Material List tab:

   • For Earth FILL, Fill Factor, enter **1.15**.

   • For Earth CUT, Cut Factor, enter **1.15**.

   • For Earth CUT, Refill Factor, enter **0.9**.

   • Click OK.

Next, you review the quantity takeoff criteria for the pavement structure.

4   On the Settings tab, right-click Pavement Structure. Click Edit.

5   In the Quantity Takeoff Criteria - Pavement Structure dialog box, Material List tab:

- Expand the trees. Review the settings.

- Click OK.

Next, you compute the earth cut and fill volumes.

6   On the ribbon, click the Analyze tab > Volumes and Materials panel > Compute Materials.

7   In the Select a Sample Line Group dialog box, review the settings. Click OK.

The names of the Civil 3D objects required for quantity calculations do not match the names in the quantity takeoff criteria. Next, you assign the Civil 3D® objects required for quantity calculation to the names in the criteria.

8   In the Compute Materials - SL Collection - 1 dialog box:

- For Quantity Takeoff Criteria, click Cut and Fill.

- For EG, click in the Object Name column. Select Existing Ground from the list.

- For DATUM, click in the Object Name column. Select 8th Avenue DATUM from the list.

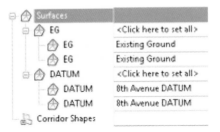

- Click OK to compute the Earth Cut and Earth Fill materials.

Next, you compute the quantities for the pavement structure using the Pavement Structure quantity takeoff criteria.

9   On the Analyze tab, click the Volumes and Materials panel > Compute Materials.

10 In the Select a Sample Line Group dialog box, click OK.

11 In the Edit Material List dialog box, notice Material List - (1), the Earth CUT and Earth FILL
    materials created for the sample line group.

    Next, you use another criteria to compute quantities.

12 In the Edit Material List dialog box, click Import Another Criteria.

13 In the Select a Quantity Takeoff Criteria dialog box, select Pavement Structure. Click OK.

    Next, you assign the object names in the drawing to the object names in the criteria.

14 In the Compute Materials - SL Collection - 1 dialog box:

    • Under Corridor Shapes, For Pave1, click 8th Avenue Pave1.

    • For Pave2, click 8th Avenue Pave2.

    • For Base, click 8th Avenue Base.

    • For SubBase, click 8th Avenue SubBase.

    Standard names are assigned to subassembly shapes. For pavement structure, the convention
    is Pave 1, Pave 2, Base, and Subbase. The name in the criteria is a generic reference to the
    material name.

    • Click OK to compute the materials. Civil 3D computes the pavement structure quantities
      and creates a second material list.

    Next, you rename the material lists.

15   In the Edit Material List - SL Collection - 1 dialog box:

- Rename Material List - (1) to **Earth Cut and Fill**.

- Rename Material List - (2) to **Pavement Structure**.

- Click OK.

Next, you review the Sample Line Group properties.

16   In the drawing area, select any sample line, right-click, and select Sample Line
      Group Properties

17   In the Sample Line Group Properties - SL Collection - 1 dialog box, click the Sections tab.

Notice that material section data has been added to the Sections list. You can display the
material section data in the section views.

18   On the Material List tab:

- Note that the material list is now attached to the sample line group.

- Scroll to the right and note the assignment of a shape style for the different materials.

- Click OK.

19   Close the drawing. Do not save the changes.

Lesson 48 | Creating Quantity Reports

This lesson describes how you create quantity reports from a sample line group. You can create quantity reports in a drawing as a table, or you can extract quantities to an external file.

After a quantity report is created, if the corridor model changes, the section data attached to the sample lines also updates. When the section data attached to the sample lines updates, the quantity table in the drawing also updates. This makes it very easy to quickly analyze quantities and adjust corridor models to balance earth cut and fill volumes.

The following illustration shows a portion of a quantity report in a table in a drawing area.

Total Volume Table						
Station	Fill Area	Cut Area	Fill Volume	Cut Volume	Cumulative Fill Vol	Cumulative Cut Vol
22+00.00	22.49	132.73	31.34	325.98	125.78	3360.02
22+50.00	44.54	84.68	74.13	219.93	199.91	3579.96
23+00.00	89.36	54.86	147.60	140.70	347.50	3720.66
23+50.00	173.73	11.25	288.62	66.40	636.13	3787.06
24+00.00	196.97	0.00	402.83	11.36	1038.96	3798.41
24+50.00	190.12	0.00	416.84	0.00	1455.80	3798.41
25+00.00	100.86	17.09	311.71	17.90	1767.51	3816.31
25+50.00	46.68	38.46	157.09	59.15	1924.61	3875.46
26+00.00	14.99	60.49	65.66	105.36	1990.27	3980.83
26+50.00	1.18	95.42	17.22	166.02	2007.49	4146.85

Objectives

After completing this lesson, you will be able to:

- Describe quantity reports.

- Create quantity reports that display quantity calculations.

- Create a quantity report in a table and a quantity report in a web browser.

About Quantity Reports

Quantity reports provide useful information to transportation designers. A common use for a quantify report is to determine how much material must be removed or added to construct the roadway. Quantity reports also form the basis for construction contracts. For earth cut and fill quantity reports, an existing surface is compared to a corridor datum surface (which represents the subgrade of the corridor model), and the results are broken down by station in a report.

Definition of Quantity Reports

Quantity reports display the total volume of material required to create a finished grade surface. The values displayed include area and volume for both cut and fill, or other specified criteria. There are default criteria and table styles for this kind of report.

Report Types

Quantity reports are either static or dynamic.

- **Dynamic**
  Dynamic quantity reports are created in the drawing area as quantity tables. Dynamic quantity reports are useful when you need to achieve a quantity balance, or generate a material excess or deficit.

- **Static**
  Static quantity reports are generated in a web browser. The format of the quantity report is controlled with a style sheet.

Example

A portion of a static quantity report generated in a web browser is shown in the following illustration.

Area Type	Area	Inc.Vol.	Cum.Vol.	MassHaul
	Sq.ft.	Cu.yd.	Cu.yd.	Cu.yd.
Station: 17+00.000				
Adjusted Cut	141.33	227.16	303.82	
Adjusted Usable	141.33	197.53	264.19	
Adjusted Fill	0.00	0.00	0.00	
				303.82
Station: 17+50.000				
Adjusted Cut	140.23	299.81	603.63	
Adjusted Usable	140.23	260.71	524.90	
Adjusted Fill	0.00	0.00	0.00	
				603.63

Creating Quantity Reports

You create quantity reports after the material quantities are calculated. You can create quantity reports in a web browser or as a table in the drawing. To generate quantity takeoff, you set criteria, associate surfaces with criteria, calculate volumes, and generate reports.

Creating Quantity Reports

To generate quantity takeoff, you set criteria, associate surfaces with criteria, calculate volumes, and generate reports.

Process: Generating a Volume Table

The following steps outline the process for generating a volume table.

1 Execute the appropriate command from the ribbon.

   • Total Volume Table creates a table in the drawing.

   • Volume Report creates a table in a web browser.

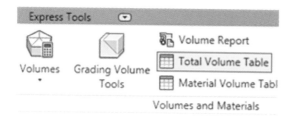

2   Pick a location in the drawing to create the table.

- Select the material list.

- Create the table.

			Total Volume Table		
Station	Fill Area	Cut Area	Fill Volume	Cut Volume	Cumulative
22+00.00	20.70	152.65	28.71	369.84	141.28
22+50.00	32.71	125.22	58.88	295.68	198.14
23+00.00	39.14	113.62	76.51	254.33	274.65
23+50.00	66.43	70.54	112.40	196.11	387.05
24+00.00	47.06	72.22	120.84	152.02	507.88
24+50.00	23.69	100.08	75.34	183.45	583.22
25+00.00	1.15	180.64	28.45	277.60	609.88
25+50.00	0.50	179.37	1.73	352.05	611.43
26+00.00	2.17	182.41	2.84	383.94	614.27
26+50.00	1.26	137.80	3.65	319.67	617.92

Guideline

Quantity reports are created for sample lines in a sample line group. You must create multiple sample line groups when cross sections change significantly (rural to urban and vice versa, either side of bridge abutments, and so on).

Exercise 01 | Create Quantity Reports

In this exercise, you create a quantity report in a table and a quantity report in a web browser.

Area Type	Area	Inc.'	Cum.Vol
	Sq.ft.	Cu.yc	Cu.yd.
Top Asphalt	4.48	0.00	0.00
Bottom Asphalt	4.48	0.00	0.00
Top Granular	17.98	0.00	0.00
Bottom Granular	42.00	0.00	0.00
Top Asphalt	4.47	8.29	8.29
Bottom Asphalt	4.47	8.29	8.29
Top Granular	17.93	33.25	33.25
Bottom Granular	41.84	77.63	77.63

The completed exercise

1. Open *Transportation - Sections and Quantities\I_quantity_reports.dwg* (*M_quantity_reports.dwg*).

2. On the ribbon, click the Analyze tab > Volumes and Materials panel > Total Volume Table.

3. In the Create Total Volume Table dialog box:

   - For Select Material List, select Earth CUT and FILL.

   - Review the remaining settings.

   - Click OK.

4. When prompted to Select Upper Left Corner, select a location in the drawing to create the table.

15+00.00	0.00	0.00	0.00
15+50.00	0.00	0.00	0.00
16+00.00	0.00	0.00	0.00
16+50.00	0.00	125.72	0.00
17+00.00	0.00	211.94	0.00
17+50.00	0.00	176.07	0.00
18+00.00	0.00	188.47	0.00

Notice that many of the values in the lower station ranges are o. This is because sample lines were created beyond the limits of the corridor. Next, you delete the unnecessary sample lines.

5   In Prospector, expand Alignments, Centerline Alignments, 8th Avenue, Sample Line Groups, and SL Collection - 1. Click Sample Lines.

6   In the item view area of Prospector:

- Press SHIFT+Select row 2+00' (9+760) to row 16+00 (10+180).

- With the stations selected, right-click. Click Delete.

7   In the AutoCAD Civil 3D 2009 dialog box, click Yes to confirm the deletion.

The sample lines are removed from the drawing and the table updates. Recall that the quantity table is dynamic.

Total Volume Table					
Station	Fill Area	Cut Area	Fill Volume	Cut Volume	Cumulativ
16+50.00	0.00	125.72	0.00	133.87	0.00
17+00.00	0.00	211.94	0.00	359.55	0.00
17+50.00	0.00	176.07	0.00	413.16	0.00
18+00.00	0.00	188.47	0.00	388.17	0.00
18+50.00	0.00	120.58	0.00	328.71	0.00
19+00.00	0.00	113.13	0.00	244.45	0.00
19+50.00	0.36	165.02	0.40	287.98	0.40
20+00.00	16.55	116.72	18.78	288.63	19.18
20+50.00	25.02	136.37	46.15	256.65	65.33
21+00.00	7.80	197.77	36.48	338.95	101.79
21+50.00	8.28	194.88	15.65	398.98	117.45
22+00.00	20.70	152.65	30.02	352.90	147.46

8   In Prospector:

  - Collapse Alignments.

  - Expand Corridors.

  - Right-click 8th Avenue. Click Rebuild - Automatic.

9   On the command line, enter **VPORTS**. Press ENTER.

10 In the Viewports dialog box, click Two: Horizontal. Click OK.

11 Click in the top viewport.

12 In Prospector:

  - Expand Alignments and Profile Views.

  - Right-click 8th Avenue PV1. Click Zoom To.

13 Click in the bottom viewport. Zoom in to the last few rows and columns of the table.

Table

ne	Cumulative Fill Vol	Cumulative Cut Vol
	839.61	6008.97
	841.12	6321.79
	841.81	6697.05
	642.33	7156.60
	642.33	7656.07
	642.33	8142.08

14 In the top viewport:

  - Select the layout profile.

  - Select a PVI near the end of the profile view on the right (this is where the corridor is calculated).

  - Move the mouse to a new location nearby and click.

The corridor rebuilds, and the quantity table automatically updates.

Table		
ne	Cumulative Fill Vol	Cumulative Cut Vol
	1100.47	4965.55
	1210.53	5046.10
	1296.52	5181.47
	1331.84	5403.77
	1335.40	5717.42
	1336.68	6110.89

Next, you create a static (nonreactive) report in a web browser.

15   On the command line, enter **VPORTS**. Press ENTER.

16   In the Viewports dialog box, select Single. Click OK.

17   On the ribbon, Analyze tab, Volumes and Materials panel, click Volume Report.

18   In the Report Quantities dialog box:

- For Select Materials List, select Pavement Structure.

- For Select a Style Sheet, select File Folder.

- Click Select Material.xsl. Click Open.

- Click OK. A web browser is launched.

19   If you are using Internet Explorer, click Yes. A portion of the report is shown.

Area Type	Area	Inc.'	Cum.Vol
	Sq.ft.	Cu.yd	Cu.yd.
Top Asphalt	4.48	0.00	0.00
Bottom Asphalt	4.48	0.00	0.00
Top Granular	17.98	0.00	0.00
Bottom Granular	42.00	0.00	0.00
Top Asphalt	4.47	8.29	8.29
Bottom Asphalt	4.47	8.29	8.29
Top Granular	17.93	33.25	33.25
Bottom Granular	41.84	77.63	77.63

20   Close the web browser. Close the drawing. Do not save the changes.

Lesson 49 | Creating Section Views

This lesson describes how you create section views.

You use section views to display surface, corridor surface, corridor, pipe network, and material section data at the sample line locations. Section data in the section views is automatically updated when the corridor recalculates or section data changes.

Objectives

After completing this lesson, you will be able to:

- Describe section views.

- Create section views from sample lines.

- Create multiple section views.

About Section Views

After you create sample lines, you create section views to show section data at the sample line locations.

Definition of Section Views

Section views are the grid objects that display section data. Section views are dynamic and are updated automatically if the geometry or location of the sample lines change. The section view style controls the display of the section view.

The following illustration shows a section view with corridor section data, existing ground surface section data, and material section data.

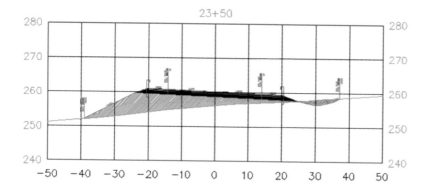

Section View Style and Groups

A section view style controls the display of section views. The section data that you display in a section view originates from the sample lines in the sample line group.

Section views are organized in a section view group, which is shown as a collection under the sample line group in Prospector. You can modify the properties of a section view group to:

- Change the group plot style.

- Change the style assigned to section views.

- Change the volume tables attached to section views.

- Show profile grade lines for other alignments and profiles in section views.

Example

The following illustration shows a section view created using a sample line extending 100 feet in each direction from the alignment centerline. The top label identifies the station where the sample line intersects the alignment. The section line has labels showing the offset and elevation values at 10-foot intervals. A data band marking the offset distance is included at the bottom of the graph.

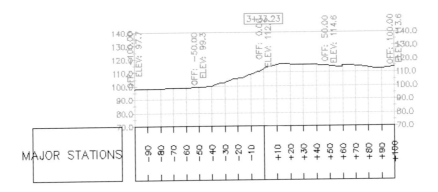

Creating Section Views

After you create sample lines, you create section views to show the section data at different stations along an alignment.

Creating Section Views

After you create sample lines, you can either create a single or multiple section views. To create multiple section views select Sections - Create Multiple Section Views from the menu. When you create multiple section views, you specify the following in the Create Multiple Section View wizard:

- Alignment and sample line group.

- Station range for section views to create.

- Section view style.

- Group plot style.

- Offset and elevation ranges.

- The section data and associated styles to view in the section views.

- Material table details and placement relative to the section views.

The section views are then plotted in the drawing.

Section views are shown in the following illustration:

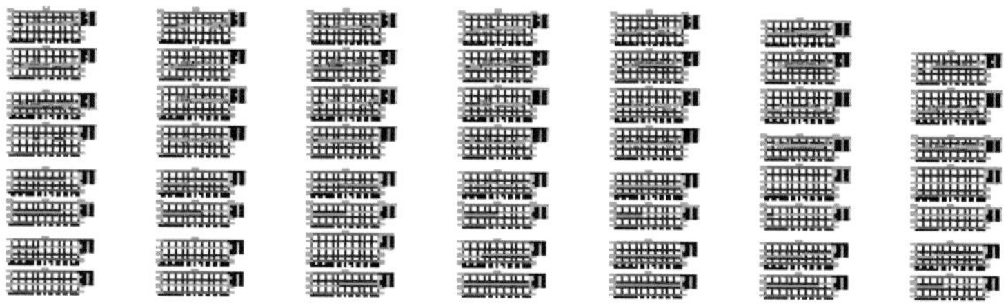

To erase section views, you delete the section view group in Prospector.

Exercise 01 | Create Multiple Section Views

In this exercise, you create multiple section views.

The completed exercise

1 Open *\Transportation - Sections and Quantities\I_section_views.dwg (M_section_views.dwg)*.

2 On the ribbon, click Home tab > Profilen and Section Views panel > Create Multiple Views.

3 In the Create Multiple Section Views wizard, General page, review all the settings. Click Next.

4 On the Offset Range page, review the settings. Click Next.

5 On the Elevation Range page, review the settings. Click Next.

6 On the Section Display Options page:

- Clear the following check boxes: Top Asphalt, Bottom Asphalt, Top Granular, and Bottom Granular.

- Existing Ground ☑
- 8th Avenue ☑
- 8th Avenue DAT... ☑
- 8th Avenue TOP ☑
- Earth CUT ☑
- Earth FILL ☑
- Top Asphalt ☐
- Bottom Asphalt ☐
- Top Granular ☐
- Bottom Granular ☐

- Click Next.

7   On the Data Bands page, review the settings. Click Next.

8   On the Section View Tables page:

- For Type, click Total Volume.

- For Select Table Style, click Standard.

- Click Add.

- For X Offset enter **0.5" (15** mm).

- Click Create Section Views.

9   When prompted to Identify Section View Origin, select a location in the drawing above the 8th Avenue Profile View and click.

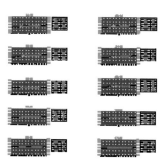

10 In Prospector, expand Alignments, Centerline Alignments, 8th Avenue, Sample Line Groups, SL Collection - 1, and Section View Groups. Click Section View Group - 1.

The section views are displayed in the Item View area.

11 In the Item View area, right-click 23+50.00 (10+420). Click Zoom To. The drawing area navigates to section view. Notice the cut and fill shapes.

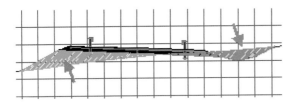

Next, you change the annotation on the cross sections. This is controlled with a code set style.

12 In Prospector, right-click SL Collection - 1. Click Properties.

13 In the Sample Line Group Properties - SL Collection - 1 dialog box, Sections tab:

• For 8th Avenue, click in the Style column.

• In the Code Set Style dialog box, select Codes with Labels.

• Click OK to close both dialog boxes.

Slope labels are applied to the corridor surface, and elevation and offset labels are applied to the corridor section points.

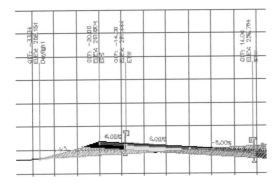

14   The multiple section views appear as follows:

15   Close the drawing. Do not save the changes.

Chapter 14
Manage Data

Data management is a very important and challenging aspect of civil engineering design projects. Large civil engineering projects that involve input from many design team members often involve changes during the design process. These changes need to be shared with other design team members in a way that eliminates the duplication of data. This chapter describes how Civil 3D® tools such as plan production, data shortcuts, quantity takeoff, and Autodesk® Vault can help the design process by offering designers the necessary tools for efficient data sharing and management.

In the first lesson, you use plan production tools to automate the layout and creation of plan and profile construction sheets. When you create sheets using the plan production tools, you first use page setup definitions in a drawing template (DWT) file to layout the sheets. You then create the final sheet layouts in either the current drawing or in new drawings. When you create sheets in new drawings, the drawing and alignment/profile object data is referenced from the original source drawings. External references are also automatically incorporated to share the graphical data. This means that changes in an engineering design drawing are automatically reflected in the production drawings. Drawings created using the plan production tools are managed using the AutoCAD Sheet Set Manager.

In the second lesson, you share drawing object data using data shortcuts and reference objects. Object data in a source drawing can be created in a reference drawing using data shortcuts. Changes to the object data in the source drawing are automatically synchronized in the reference drawings. This functionality is useful when designs are created in design drawings, and the object data is referenced in production drawings used for creating plan and profile construction sheets.

In the third lesson, you perform a quantity takeoff and create a report that summarizes the quantities in the drawing. Quantities are summarized and reported for AutoCAD® entities such as blocks, polylines, and closed polygons. Quantities can also be summarized and reported for Civil 3D pipe networks and corridor models.

In the final lesson, you share object and drawing data using Autodesk Vault. Autodesk Vault is a data management and sharing tool that enables you to efficiently share object, drawing, and image data among several design team members. Autodesk Vault also incorporates data versioning and provides read/write access control. You check data in and out of the vault to work with the most recent version of that data.

Objectives

After completing this chapter, you will be able to:

- Use plan production tools to automate the process of creating construction documents from your designs.

- Share drawing data using data shortcuts and reference objects.

- Calculate quantity takeoff using assigned pay items.

- Check in and check out Civil 3D files from Vault.

Lesson 50 | Plan Production

This lesson describes how to automate the generation of plan and profile design and construction sheets using the plan production tools. These tools, the Create View Frames Wizard and Create Sheets Wizard, eliminate the repetitive tasks associated with orienting and scaling viewports to show alignment and profile data.

Using the wizards, you can quickly create sheets that display segments of alignments and profiles in your design and construction plans. Instead of having to manually create viewports for alignments and profile views, and manually recreate sheets each time your data changes, you can now create sheets from dynamic view frame groups that automatically capture predefined areas along an alignment and a profile view.

Objectives

After completing this lesson, you will be able to:

- Describe the Create View Frames wizard pages and list the guidelines for using the wizard.

- Describe the Create Sheets wizard pages and list the guidelines for using the wizard.

- Use the Create View Frames and Create Sheets wizards to create a construction plan.

Creating View Frames

You use the Create View Frames wizard to quickly create view frames along an alignment.

Create View Frames Wizard Pages

Using the following areas of the Create View Frames wizard, you can plan and create your sheet sets.

Term	Description
Alignment	Select the alignment and station range for creating view frames.
Sheets	Select the type of sheets to create, a template for the sheets, and to determine view frame placement.
View Frame Group	Specify criteria for creating the view frame group object.
Match Lines	Configure a variety of choices that determine how and if match lines will be placed on the view frames.
Profile Views	Choose options for the profile views that will be displayed in the viewports (sheets).

The Create View Frames wizard is shown in the following illustration.

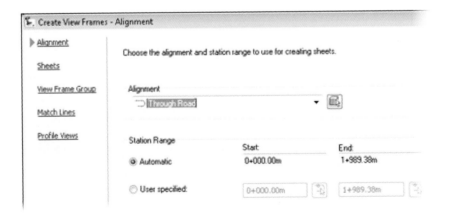

Guidelines

Keep the following guidelines in mind when using the Create View Frames wizard.

- You can create view frames automatically using the Create View Frames wizard.

- Before you create view frames, the desired alignment must already exist in your drawing.

- View frames represent rectangular areas along the alignment that will be displayed in sheets or construction documents.

- Depending on the type of sheets you want to produce (plan and profile or profile only), you may also need to have a profile already created.

- If you are creating a plan only view frame or sheet set, then you do not need to have a profile in the drawing.

- The profile view style and band set that you specify in the Create View Frames wizard are used to calculate the placement of view frames. You cannot change these styles later for the view frame group.

Creating Sheets

You can automatically create sheets for construction documents or plans using the Create Sheets wizard. Before you can use the Create Sheets wizard, you must have already created a view frame group in the drawing, using the Create View Frames wizard.

Create Sheets Wizard

You can use the following areas of the Create Sheets wizard to plan and create your sheet sets.

Term	Description
View Frame Group and Layout	Choose the view frame group and output settings for layout creation.
Sheet Set	Specify creation criteria for the sheet set, such as names and locations for the sheet set, the sheet set file (DST), and the sheet file.
Profile Views	View the profile view style and band set chosen during view frame creation, configure profile view options.
Data References	Select the objects to be referenced in sheets.

The Create Sheets wizard is shown in the following illustration.

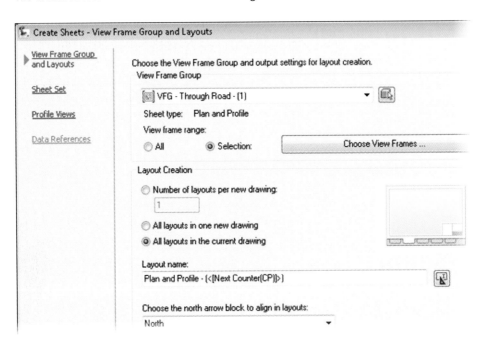

Guidelines

Keep the following guidelines in mind when working with the Create Sheets wizard.

- You cannot change the profile view style and band set in the Create Sheets wizard. If you want to change the profile view style or band set, click Cancel on the Create Sheets wizard. Return to the Create View Frames wizard, change the profile view style and or the band set there, recreate the view frames, and restart the Create Sheets wizard.

- If you have chosen an output that does not include profiles, the Profile View section of the Create Sheets wizard is not displayed.

- If you have chosen to have the sheets saved in the current file, then the Data References page of the Create Sheets wizard is bypassed.

Exercise 01 | Create Construction Plans

In this exercise, you use the Create View Frames and Create Sheets wizards to create a project construction plan.

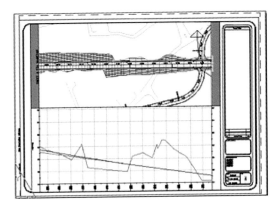

The completed exercise

Create View Frames

First, you use the Create View Frames wizard to generate frames for use in creating plan and profile sheets.

1   Open...*Manage Data\create_view_frames.dwg.*

2   On the ribbon, click the Output tab. On the Plan Production Panel, click Create View Frames.

3   In the Create View Frames dialog box, Alignment page:

- For Alignment, select Through Road.

- For Station Range, click Automatic.

- Click Next.

4   On the Sheets page:

- Under Sheet Settings, for Sheet Type, click Plan and Profile.

- For Template for Plan and Profile sheet, click the ellipsis.

5   In the Select Layout as Sheet Template dialog box:

- Click the ellipsis.

- In the Select Layout as Sheet Template dialog box, navigate to where you installed the Creating Sheet Sets data set.

- Select *Civil 3D (Metric) Plan and Profile.dwt*. Click Open.

- Under Select a Layout to Create New Sheets, select Plan and Profile 1 to 500.

- Click OK.

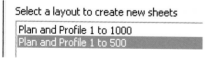

6   On the Sheets page:

- Under View Frame Placement, click Along Alignment.

- Select the Set the First View Frame Before the Start of the Alignment By check box.

- Enter **5**.

- Click Next.

> **View Frame Placement**
> - ⦿ Along alignment
> - ○ Rotate to north
> - ☑ Set the first view frame before the start of the alignment by:
>
>   5.000m

7   On the View Frame Group page:

- For Label Location, select Top Center.

- Click Next.

8   On the Match Lines page, under Positioning:

- Select the Snap Station Value Down to the Nearest check box.

- Enter **1**.

- Select the Allow Additional Distance for Repositioning check box.

- Enter **3.00** m.

- For both the Left and Right Label Locations, select Middle.

- Click Next.

9   On the Profile Views page:

- For Profile View Style, select Major Grids.

- For Select Band Set Style, select EG-FG Elevations and Stations.

- Click Create View Frames.

The view frames, matchlines, and associated labels are created along the Through Road alignment.

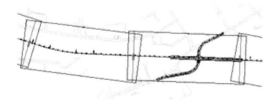

Create Sheets

Next, you use the Create Sheets wizard to create sheets and sheet set files.

1   Open...*Manage Data\create_sheets.dwg*.

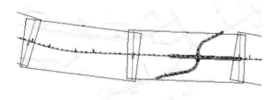

2   On the ribbon, Output tab, Plan Production Panel, click Create Sheets.

3   On the View Frame Group and Layouts page, under View Frame Group:

   • For View Frame Range, click Selection.

   • Click Choose View Frames.

4   In the Select View Frames dialog box:

   • Press CTRL+select the first four view frames.

   • Click OK.

5   On the View Frame Group and Layouts page, under Layout Creation:

   • Click All Layouts in the Current Drawing.

   • Change the Layout Name to **Plan and Profile - (<[Next Counter(CP)]>)**.

   • For Choose the North Arrow Block to Align in Layouts, select North.

   • Click Next.

○ All layouts in the current drawing

Layout name:

Plan and Profile - (<[Next Counter(CP)]>)

Choose the north arrow block to align in layouts:

North ▼

6   On the Sheet Set page, under Sheet Set:

- Click New Sheet Set.

- Verify that the Sheet Set File (.dst) Storage Location is set to where you installed the Civil 3D Essentials 2010 data directory.

- Click Next.

7   On the Profile Views page, under Other Profile View Options:

- Click Choose Settings.

- Click Profile View Wizard.

8   In the Create Multiple Profile Views wizard:

- Click Next as you verify the settings for each page in the wizard.

- Click Finish.

9   On the Profile Views page:

- Click Create Sheets.

- Click OK to save the drawing.

10   When prompted to Select Profile View Origin, select a location in the drawing area above the plan and click.

Acknowledge the messages in the Event viewer (Panorama window).

11   In the Sheet Set Manager dialog box, double-click each sheet to review it.

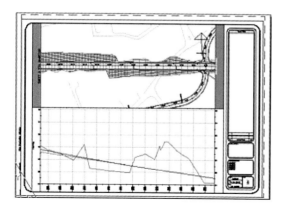

12   Close the drawing. Do not save changes.

Lesson 51 | Working with Data Shortcuts and Reference Objects

This lesson describes how you use data shortcuts and reference objects to share drawing object data. When you share drawing object data using data shortcuts and reference objects, you eliminate the errors associated with copying and transposing data between drawings.
Data shortcuts and reference objects are powerful drawing object data sharing mechanisms that enable several members of a design team to simultaneously work on projects and reference the same data sources.

Objectives

After completing this lesson, you will be able to:

- Explain the function of data shortcuts and reference objects.

- Describe the process for working with data shortcuts and reference objects.

- Work with data shortcuts and reference objects.

About Data Shortcuts and Reference Objects

Data shortcuts and reference objects enable you to share drawing object data that exists in one drawing with several other people working in their own drawings.

Definition of Data Shortcuts

Residing in XML files, data shortcuts are external links to drawing object data. You create a data shortcut for an object in one drawing so you can reference that same object in a second drawing.

Definition of Reference Objects

Reference objects are objects created in a drawing that reference object data from a source drawing. Reference objects are created from data shortcuts and cannot be edited. However, they can be displayed and queried like regular objects.

Additional Information

- Data shortcut and reference object functionality supports surface, alignment, profile, pipe network, and plan production view frame groups.

- Reference object data is read-only and cannot be changed. However, the user can modify the properties of the reference object and apply a local object style, apply custom annotation, or perform limited analysis to the referenced data.

- The reference object takes up less file space when the consumer drawing is saved.

- If you create a reference to a profile data shortcut, a data shortcut to the profile's parent alignment is also created.

- You can synchronize reference objects with their source drawings. When source drawing object data changes, users referencing this data are notified that the object data has changed. You then have the option to synchronize the reference data with the source data.

- You use the Data Shortcuts manager to set the working folder for the data shortcuts, create data shortcuts, export the data shortcuts to a file, import the data shortcuts from a file, and create the reference objects.

Example

An engineer chooses to design a network of subdivision roads and utilities in a single engineering design drawing. Data shortcuts for the design alignments, profiles, pipe networks, and surfaces are created. Another design team member then creates reference objects with annotation from the data shortcuts in plan and profile production drawings. When the design changes in the engineering drawings, the production drawings are synchronized and automatically updated.

Working with Data Shortcuts and Reference Objects

Before you create data shortcuts, you first create an active data shortcuts project, or a working folder. When you create data shortcuts, you create a separate XML file for each selected surface, alignment, or other eligible object. These XML files are stored in subfolders of the _Shortcuts folder for the active data shortcuts project. The XML files are used internally to create data references to the source object in other drawings.

Using data shortcuts involves two main tasks:

- Create data shortcuts from the source drawing.
- Create reference objects in the reference or consumer drawing.

Working with Data Shortcuts and Reference Objects

The process for working with data shortcuts involves working with the source drawing, which contains the object data, and with the reference drawing, which is where the reference objects are created.

The steps for working with data shortcuts and reference objects are as follows:

- Create the *Data Shortcuts* folder. This is the location where the data shortcut files will be located.

- From the source drawing, create the data shortcut. This results in the creation of the data shortcut XML file in the data shortcut project folder.

- From the reference drawing, create the reference object. This results in the creation of an object that references the data in the source drawing.

Guidelines

Keep the following guidelines in mind when working with data shortcuts.

- The default working folder for data shortcut projects is *C:\Civil 3D Projects*. Change the working folder to your company project folder to better organize the data.

- When you create a new data shortcut folder, specify a name that reflects the project name. Specify a project template, the standard structure of folders and subfolders for keeping project data organized, that conforms to company standards.

- To control access to data shortcuts, remove the Data Shortcuts node from Prospector. Designers permitted to use data shortcuts enter the ShortCutNode command to either hide (disable) or show (enable) Data Shortcuts.

Exercise 01 | Work with Data Shortcuts and Reference Objects

In this exercise, you learn how to create data shortcuts for object data in one drawing, and create reference objects from the data shortcuts in a second drawing.

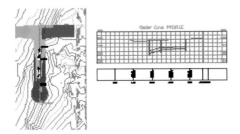

The completed exercise

1   Open \Manage Data\data_shortcuts_reference_objects.dwg.

First, you create a folder for the data shortcuts.

2   In Prospector, change the view to Master View.

3   Right-click Data Shortcuts. Click New Data Shortcuts Folder.

4   In the New Data Shortcut Folder dialog box, for Name, enter **C3D 2010 AOTG Shortcuts**. Click OK.

A folder called *C3D 2010 AOTG Shortcuts* is created in Civil 3D Projects. This is where the data shortcut files are organized.

5   On the ribbon, Manage tab, Data Shortcuts panel, click Create Data Shortcuts.

You create shortcuts for all surface, alignments, profiles, and pipe networks.

6   In the Create Data Shortcuts dialog box, select the check boxes for the following objects:

- Surfaces

- Alignments

- Pipe Networks

- Click OK.

Next, you review the data shortcut files.

7   In Prospector, expand Data Shortcuts.

Note the data shortcuts that you just created.

8   Launch Windows Explorer.

9   In Explorer:

- Browse to ...\Civil 3D Projects\C3D 2010 AOTC Shortcuts.

- Review the contents of the folders.

- Notice the data shortcut XML files.

- Close Windows Explorer.

10 In Civil 3D, save the drawing.

Next, you create a new drawing and create the reference objects.

11 On the Application menu, click New.

12 In the Select Template dialog box:

- Click _AutoCAD Civil 3D (Imperial) NCS.dwt.

- Click Open.

13 On the Application menu, click Save As.

14 In the Save Drawing As dialog box:

- Browse to ...*Manage Data*.

- For File Name, enter **reference_objects**.

- Click Save.

15 In Prospector:

- Change the view to Master View.

- Expand Data Shortcuts (*C:Civil 3D ProjectsC3D 2010 AOTC Shortcuts*).

- Expand Surfaces, Alignments, Pipe Networks. Note the data shortcuts.

- Under Surfaces, right-click Existing Ground. Click Create Reference.

16 In the Create Surface Reference dialog box:

- For Style, click in the Value column. Click the ellipsis.

- Select Contours 1' and 5' (Background) from the list.

- Click OK twice to close the dialog boxes.

17 Zoom to the extents of the drawing.

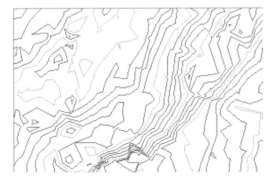

18   In Prospector, expand Surfaces and Existing Ground.

Notice the referenced surface object named Existing Ground. There is no Definition tree because the data for the reference object cannot be modified.

19   In Prospector, under Data Shortcuts, Surfaces, right-click FG Roads Top. Click Create Reference.

20   In the Create Surface Reference dialog box:

- For Style, click in the Value column. Click the ellipsis.

- Select Elevation Banding (2D) from the list.

- Click OK twice to close the dialog boxes.

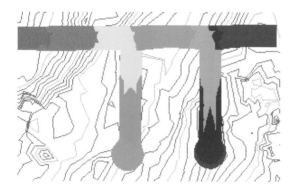

21   In Prospector, under Data Shortcuts, Alignments, right-click Cedar Cove. Click Create Reference.

22   In the Create Alignment Reference dialog box:

- For Alignment Label Set, click Major Minor and Geometry Points.

- Click OK.

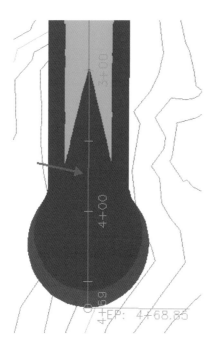

23   In Prospector:

- Expand Data Shortcuts, Alignments, Cedar Cove, Profiles.

- Right-click Cedar Cove FG. Click Create Reference.

24   In the Create Profile Reference dialog box:

- For Profile Style, click Design Profile.

- For Profile Label Set, click Complete Label Set.

- Click OK.

25   In Prospector:

- Expand Data Shortcuts, Pipe Networks, Storm 1.

- Right-click Storm 1. Click Create Reference.

26 In the Create Pipe Network Reference dialog box:

- Review the contents.

- Click OK.

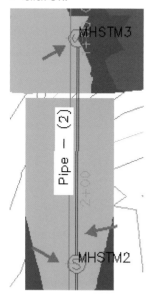

Next, you create a profile view for the Cedar Cove alignment.

27 On the ribbon, Home tab, Profile & Section Views panel, click Profile View > Create Profile View.

28 In the Create Profile View wizard:

- On the General page, for Profile View Style, click Profile View. Click Next.

- On the Station Range page, click Next.

- On the Profile View Height page, click User Specified.

- For Minimum, enter **284**.

- For Maximum, enter **300**.

- Clear the Split Profile View check box. Click Next.

- On the Profile Display Options page, click Next.

- On the Pipe Network Display page, click Storm 1.

- Click Create Profile View.

29 When prompted to Select Profile View Origin, click a location in the drawing to the right of the plan view.

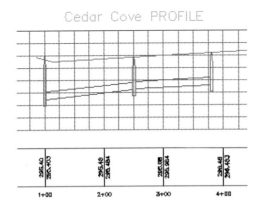

30 Save the drawing.

31 In Prospector, right click *data_shortcuts_reference_objects.dwg*. Click Switch To.

32 In the drawing area:

- Zoom to Cedar Cove in plan view.

- Click the north structure of the pipe network.

- Click the square grip.

- Click a location near station 2+00.

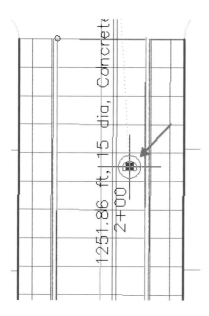

33 Save the drawing.

34 Press CTRL+TAB to switch to the *reference_objects* drawing.

You are notified on the bottom right that the data shortcut definitions may have changed.

35 In the warning dialog box, click Synchronize.

Acknowledge the messages in the Event viewer (Panorama window). The pipe network updates in both plan and profile views.

> ⓘ **Data shortcut definitions may have changed**
>
> References to data shortcut definitions may have changed and may require synchronization
>
> Synchronize

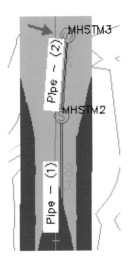

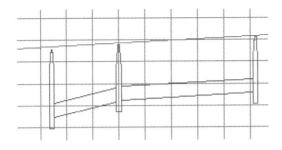

36 The completed exercise drawing appears as follows:

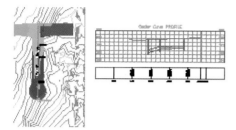

37 Close the drawings. Do not save the changes.

Lesson 52 | Calculating Quantity Takeoff Using Pay Items

This lesson describes how you use pay items to calculate quantity takeoff. You use the quantity takeoff functionality to generate a list of items, and their respective quantities, for use in project estimating or a construction contract.

The following illustration shows a quantity takeoff report.

```
 Quantity Takeoff Report

                              Summary Takeoff
                             -----------------

 Pay Item ID  Description                                Quantity   Unit
 -----------  ----------------------------------------   ---------- ----
 60201-0400   12-INCH CONCRETE PIPE                      524.377    LNFT
 60201-0500   15-INCH CONCRETE PIPE                      227.238    LNFT
 60201-0600   18-INCH CONCRETE PIPE                      230.010    LNFT
 60401-2000   MANHOLE, TYPE 2                            6          EACH
 60901-0400   CURB, CONCRETE, 6-INCH DEPTH              1507.461    LNFT
 61106-0000   FIRE HYDRANT                               8          EACH
 62401-0300   FURNISHING AND PLACING TOPSOIL, 4-II       894.13     SQYD
 62511-2000   SEEDING, HYDRAULIC METHOD                  894.13     SQYD
 62516-1000   MULCHING, DRY METHOD                       894.13     SQYD
```

Objectives

After completing this lesson, you will be able to:

- Explain the purpose of quantity takeoff.

- Describe pay items and pay item lists.

- List the entities and objects that you can assign to pay items.

- Describe a process and list guidelines for performing quantity takeoff.

- Calculate quantity takeoff using pay items.

About Quantity Takeoff

Quantity takeoffs are used in both the planning and detailed design phases of civil engineering projects. For planning projects, an engineer performs a quantity takeoff to estimate construction costs. For detailed design projects, an engineer performs a quantity takeoff to calculate the precise quantities for use in a construction contract.

Definition of Quantity Takeoff

Quantity takeoff is the act of calculating and reporting material quantities from a drawing.

Example

The illustration below shows a Summary Takeoff report. The report includes the pay item ID, pay item description, quantity, and unit. The pay item ID, description, and unit originate in the pay item list. The displayed quantity is a result of performing a quantity takeoff and represents the quantity of materials in the drawing with pay items assigned to them.

```
  Quantity Takeoff Report

                            Summary Takeoff
                            ----------------.
  Pay Item ID  Description                          Quantity   Unit
  -----------  ----------------------------------- ---------- ----
   60201-0400   12-INCH CONCRETE PIPE                524.377    LNFT
   60201-0500   15-INCH CONCRETE PIPE                227.238    LNFT
   60201-0600   18-INCH CONCRETE PIPE                230.010    LNFT
   60401-2000   MANHOLE, TYPE 2                      6          EACH
   60901-0400   CURB, CONCRETE, 6-INCH DEPTH         1507.461   LNFT
   61106-0000   FIRE HYDRANT                         8          EACH
   62401-0300   FURNISHING AND PLACING TOPSOIL, 4-II 894.13     SQYD
   62511-2000   SEEDING, HYDRAULIC METHOD            894.13     SQYD
   62516-1000   MULCHING, DRY METHOD                 894.13     SQYD
```

About Pay Items

To perform a Quantity Takeoff, you must first create and assign pay items to the appropriate objects in your drawing. This section describes pay items and assigning them to Civil 3D objects.

Definition of Pay Items

A pay item is an item in a pay item list that is referenced in either a construction tender or in a project cost estimate. The pay item consists of the item number, a description, and a unit cost. In Civil 3D, pay items are assigned to Civil 3D objects and AutoCAD entities.

Pay Item Lists

Before you can perform a quantity takeoff, you must create a pay item list. A sample pay item list is included in Civil 3D. You can copy and modify this sample to meet the requirements of your client or jurisdiction.

Pay items can be associated with AutoCAD lines, open or closed polylines, blocks, or any AutoCAD or Civil 3D entity after they have been created. You can assign pay items to a Civil 3D code set style so that corridor objects are automatically tagged with specific pay items when they are created. You can also assign pay items to a Civil 3D pipe network parts list, so that new pipes and structures are automatically tagged with the specific pay items when they are created. You can also add commonly used pay items to a Favorites list for ease of access.

Different pay item lists can be associated and saved with different drawings. A pay item file contains pay item codes, descriptions, and units of measure. The pay item file is formatted as either a Comma Separated Variable (CSF) or an eXtensible Markup Language (XML) file. An optional pay item categorization file categorizes pay items into manageable groups.

Example

The following illustration shows a pay item list. The pay item list is displayed in the Quantity Takeoff Manager in the Panorama window.

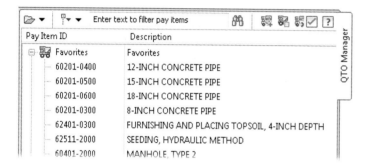

Assigning Pay Items

After you have opened the pay item list and attached it to the drawing, you assign specific pay items to AutoCAD entities and Civil 3D objects in the drawing using the QTO Manager window.

Assigning Pay Items

Pay items can either be assigned manually or automatically. You can manually assign pay items to entities and objects after they have been created. Examples of pay items you manually assign are listed in the following table.

Type	Description
Closed polyline (area)	This pay item type is typically assigned to seeding and mulching, topsoil, sod, and watering objects.
Open polyline (linear)	This pay item type is typically assigned to curb and gutter, pavement marking, and fence line objects.
Block (each)	This pay item type is typically assigned to fire hydrant, catch basin, water valve, and tree objects.

Pay items can be assigned automatically for AutoCAD Civil 3D pipe networks and corridor objects.

Assigning Pay Items to Pipe Network Parts

You can modify a pipe network parts list to assign pay items. When you create parts using a parts list that has assigned pay items, the parts are automatically tagged with the pay items.

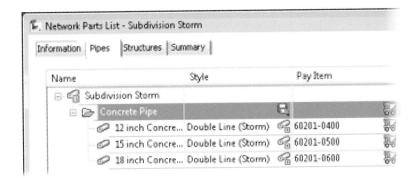

Pipe network parts list

Assigning Pay Items to Corridors

The Code Set Style can accommodate pay items for both linear and pavement structure features of a corridor model. You can assign a lineal foot pay item to the Flowline_Gutter point code in a Code Set Style. When a corridor is created using the Code Set Style, the pay item is automatically assigned to the Flowline_Gutter point codes. Performing a quantity takeoff will result in the calculation of the lineal feet of curb and gutter.

You can assign a square foot, depth-based pay item to corridor link definitions in a Code Set Style. When a corridor is created using the Code Set Style, the pay item is automatically assigned to the links. Performing a quantity takeoff will automatically calculate the volume of materials in the pavement structure.

Performing Quantity Takeoff

This section describes how you perform a quantity takeoff. You perform a quantity takeoff to generate a list of items and quantities for the drawing. When you perform a quantity takeoff, the items are summarized and presented in a table. The table can be saved to an external file. The quantity take off command is launched from the ribbon, Analyze tab, QTO panel.

QTO Manager   Takeoff

QTO

Performing Quantity Takeoff

1   Create a Pay Item List that contains the pay items for your client/jurisdiction.

2   Assign pay items to AutoCAD entities and drawing objects.

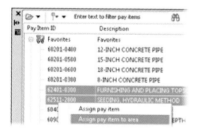

3   Perform a quantity takeoff.

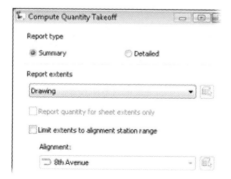

Exercise 01 | Calculate Quantity Takeoff Using Pay Items

In this exercise, you add a pay item to your favorite pay items list, you assign pay items to AutoCAD entities and Civil 3D objects in the drawing, and you perform a quantity takeoff on the objects with assigned pay items.

Pay Item ID	Description
60201-0400	12-INCH CONCRETE PIPE
60201-0500	15-INCH CONCRETE PIPE
60201-0600	18-INCH CONCRETE PIPE
60401-2000	MANHOLE, TYPE 2
60901-0400	CURB, CONCRETE, 6-INCH DEPT
61106-0000	FIRE HYDRANT
62401-0300	FURNISHING AND PLACING TOPS
62511-2000	SEEDING, HYDRAULIC METHOD
62516-1000	MULCHING, DRY METHOD

The completed exercise

1   Open \Manage Data\QTO.dwg.

2   On the ribbon, Analyze tab, QTO panel, click QTO Manager.

The QTO Manager displays in the Panorama window. Next you open the pay item file.

3   In the QTO Manager, click the file folder icon. Select Open Pay Item File.

4   In the Open Pay Item File dialog box:

- For Pay Item File, click the file folder icon. Browse to ...*Manage Data. Select Getting Started. csv*. Click Open.

- For Pay Item Categorization file, click the file folder icon. Browse to ...*Manage Data*. Select *Getting Started Categories.xml*. Click Open.

- Click OK. A pay item list is displayed in the QTO Manager.

IMAGE23

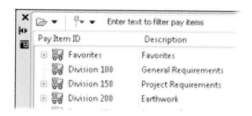

Favorites to see a list of commonly used pay items.

Next, you add an additional item to the favorites list.

6   Expand Division 600 Incidental Construction.

7   Expand Group 625 Turf Establishment.

8   Expand Section 62516 Mulching, ----(area).

9   Right-click 62516-1000, click Add to Favorites List.

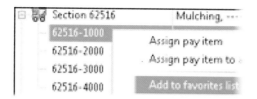

10   At the top of the pay item list, expand Favorites.

Notice the addition of pay item 62515-1000 MULCHING, DRY METHOD.

Next, you assign pay items to the boulevards where planting is to occur.

11   In QTO Manager, CTRL+select the following pay items in the Favorites list:

- 62401-0300 FURNISHING AND PLACING TOPSOIL, 4-INCH DEPTH

- 62511-2000 SEEDING, HYDRAULIC METHOD

- 62516-1000 MULCHING, DRY METHOD

62401-0300	FURNISHING AND PLACING TOPS
62511-2000	SEEDING, HYDRAULIC METHOD
60401-2000	MANHOLE, TYPE 2
60901-0400	CURB, CONCRETE, 6-INCH DEPTH
61106-0000	FIRE HYDRANT
62516-1000	MULCHING, DRY METHOD

- Right-click, click Assign Pay Item to Area.

12   At the Select Point prompt, enter **O** (for select Object). Press ENTER.

13   At the Select Object prompt, in the drawing area, click the north magenta landscaping boulevard polyline on the west side of 8th Avenue.

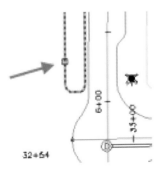

14   Click the south magenta landscaping boulevard polyline on the west side of 8th Avenue. Press ENTER.

The 3 pay items have been assigned to the landscaping boulevard polylines.

15   In the drawing area, hover the cursor over the hatching that appears in the landscaping boulevard polyline. Note the assignment of the pay items in the tooltip.

Note: You may need to move the mouse to highlight the hatching.

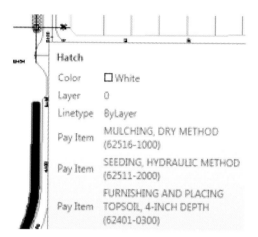

Next, you assign pay items to the fire hydrants.

16 In the drawing area, click any fire hydrant block to select it.

17 Right-click, click Select Similar to select all the fire hydrant blocks.

18 In the QTO Manager, Favorites list, right-click 61106-0000 FIRE HYDRANT; click
Assign Pay Item.

19 In the drawing area, hover the cursor over any fire hydrant. Notice the assignment of the pay
item in the tooltip.

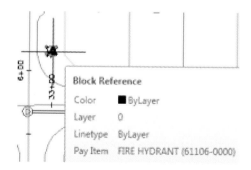

Next, you assign a pay item to the curb and gutter polylines and alignment objects.

20   In the drawing area, click the polyline representing the curb and gutter to the left of 8th Avenue.

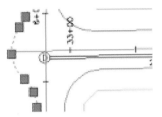

21   Right-click and click select similar.

Notice that the pavement edges adjacent to the Birch Lane and Cedar Cove cul-de-sacs are not selected. These are alignment objects.

Next, you add the alignment objects to the selection set.

22   In the drawing area, click the pavement edges adjacent to the two cul-de-sacs to select them both.

23   In the QTO Manager, under Favorites, right- click on 60901-0400 CURB, CONCRETE, 6-INCH DEPTH and click Assign Pay Item.

A curb and gutter pay item is assigned to the curb and gutter polylines and edge of pavement alignments.

24   In the drawing area, hover the cursor over any curb and gutter polyline or edge of pavement alignment and note the assignment of the pay item in the tooltip.

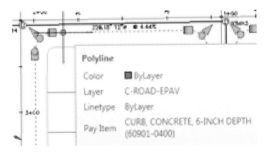

Next, you review and assign pay items to structures and pipes in a storm sewer network.

25   In Toolspace, Settings tab, expand Pipe Network and Parts Lists.

26   Right-click Subdivision Storm, click Edit.

27   In the Network Parts List dialog box, click the Pipes tab.

28   Expand Subdivision Storm and Concrete Pipe.

      Notice the Pay Item Column.

29   For 12 inch Concrete Pipe, click the Pay Item icon in the Pay Item Column.

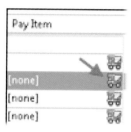

30   In the Pay Item List, expand Favorites, click 60201-0400 12-INCH CONCRETE PIPE. Click OK.

      The pay item has been assigned to the 12 inch Concrete Pipe. When you create a pipe network
      using this Parts List, the pay item is automatically assigned.

31   Click the Structures tab. Note that you can assign pay items to structures as well.

32   Click Cancel.

      The pipe network in the drawing has no pay items assigned. Next, you manually assign the
      pay items to the parts in the pipe network.

33   In the drawing area, select any manhole.

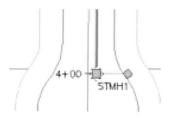

34   Right-click, click Select Similar to select all the manholes.

35   In the QTO Manager, Favorites list, right-click on 60401-2000 MANHOLE, TYPE 2.
Click Assign Pay Item.

36   In the drawing area, click the 3 x 12 inch pipe segments on Cedar Cove and on Orchard Road
between Cedar Cove and Birch Lane.

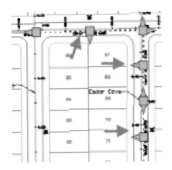

37   In the QTO Manager, Favorites list, right-click 60201-0400 12-INCH CONCRETE PIPE.
Click Assign Pay Item.

38   In the drawing area, click the 15 inch pipe segment between STMH4 and STMH5.

39   In the QTO Manager, right-click 60201-0500 15- INCH CONCRETE PIPE, click Assign Pay Item.

40   In the drawing area, click the 18 inch pipe segment between STMH5 and STMH6.

41   In the QTO Manager, right-click 60201-0600 18- INCH CONCRETE PIPE.
Click Assign Pay Item.

42   In the drawing area, hover the cursor over the structures and pipes. Review the assigned pay
items in the tool tips.

All the pay items are assigned. Next, you perform the quantity takeoff.

43   On the ribbon, Analyze tab, QTO panel, click Takeoff.

44 In the Compute Quantity Takeoff dialog box:

- For Report type, click Summary.

- For Report extents, click Drawing.

- Clear all other checkboxes.

- Click Compute.

45 In the Quantity Takeoff Report dialog box:

- On the bottom right, select Summary (HTML).xsl.

- Review the quantities and click Close.

Pay Item ID	Description
60201-0400	12-INCH CONCRETE PIPE
60201-0500	15-INCH CONCRETE PIPE
60201-0600	18-INCH CONCRETE PIPE
60401-2000	MANHOLE, TYPE 2
60901-0400	CURB, CONCRETE, 6-INCH DEPT
61106-0000	FIRE HYDRANT
62401-0300	FURNISHING AND PLACING TOPS
62511-2000	SEEDING, HYDRAULIC METHOD
62516-1000	MULCHING, DRY METHOD

46 Close the drawing. Do not save the changes.

Lesson 53 | Working with Autodesk Vault

This lesson describes how you use Autodesk Vault to share drawing and object data amongst project team members. You can add files and objects, and check files and objects in and out of the Vault. The files support DWG™ format and image types. You use Autodesk Vault 2010 Data Management Server to manage Vault projects.

Objectives

After completing this lesson, you will be able to:

- Describe Autodesk Vault.

- Describe the typical workflow for working with Civil 3D and Vault.

- Explain how you check drawings in and out of Vault.

- Manage drawings and create a reference object using Vault.

About Vault

Autodesk Vault is a complete document management system for drawings, project objects, and project-related files. Autodesk Vault provides access and version control for all project drawings.

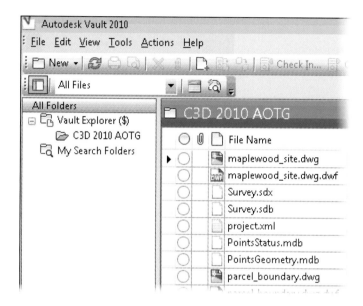

Autodesk Vault client-side window

Definition of Autodesk Vault

An Autodesk Vault database is a central repository of drawing and project data that can be shared amongst several users. Vault data can be accessed in a collection on the Projects node in Prospector.

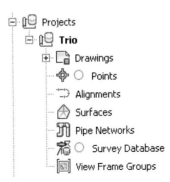

Projects in Prospector

Server-side Component

To use Autodesk Vault, you must install the Autodesk Data Management Server (ADMS). The ADMS is installed on a server and the client component is installed on each computer that needs communicate with the server. The server stores all master copies of data files, and the clients can access the files stored on the server.

Client-side Components

Project data can be accessed either from Prospector, or by using the Vault tool. The Vault tool is a stand-alone client application that enables you to access and manage Vault data.

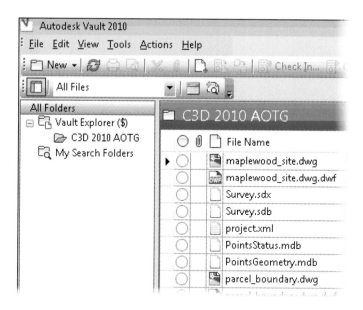

Autodesk Vault client-side window

AutoCAD Civil 3D also serves as a client for Vault data. You log into Vault from Prospector to access the data.

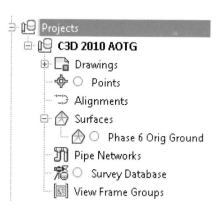

Autodesk Vault client-side window

Example of Using Vault

Using Autodesk Vault, you can share project drawings, project objects, and points with project members.

In Autodesk Vault, you create a project folder that contains the collection of project drawings. You can use folders to organize the drawings within the project. One master copy of each project drawing is maintained in a specific project within a Vault database. Only one person can edit the master copy at a time. Other team members can reference the master copy, and they are notified when the master copy is updated. Project points can be also be shared using Vault. Master copies of points are stored in a point database, and points are checked out or copied directly to a drawing. The project points in a drawing are usually a subset of the project database.

A project object can contain surfaces, alignments, profiles, pipe networks, and view frame groups that can be shared with others. All shared project objects are listed in the project object collections in the Prospector tree. Project members can create a read-only copy of a project object, called a reference, in a drawing. This read-only copy of the project object's geometry can be used to perform labeling, design, analysis, or what-if tasks. Multiple project members can create references to the same shared project object.

Working with Vault

This section describes a typical workflow using Autodesk Vault and AutoCAD Civil 3D.

Accessing Vault

The first step in working with Vault is to log in to the Vault. You can access the Vault from Civil 3D to check files in and out of the application, and keep track of attachments to files.

Procedure: Log In to Vault

The following steps describe how to access Autodesk Vault from AutoCAD Civil 3D.

1   Start AutoCAD Civil 3D.

2   In Toolspace, Prospector tab, under Master View, right-click Projects > Log In to Vault.

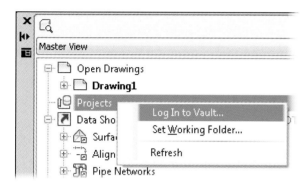

Process: Using a Typical Vault Workflow

Your workflow varies depending on where you are in the design cycle. The following general guidelines show a basic workflow for working with Vault.

1   Start Civil 3D. Log in to Vault.

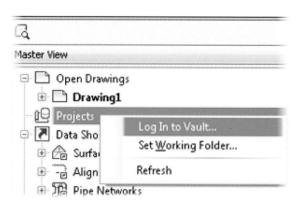

2   Create a new project. Open a drawing file you want to add to the project.

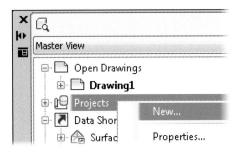

3   Add the drawing to the project.

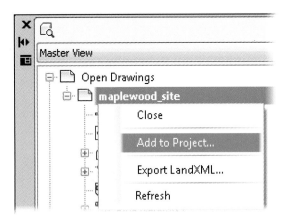

4   Check out the drawing.

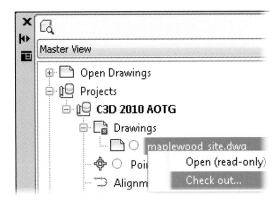

5   Modify and save the drawing. The Vault status icon changes to reflect the current state of
    the drawing.

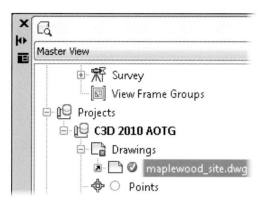

6   Check the drawing back into Vault.

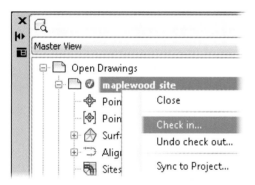

Managing Drawings

This section describes how you manage drawings in Vault. It explains how you check out and check
in drawings within a Vault project.

After you log in to Vault, the Projects object in Prospector lists all the projects that are available to
you. The Civil 3D Project object groups the drawings and data for a single project in one place and
controls access to all project files. Within each project, the Drawings, Points, Alignments, Surfaces,
Pipe Networks, and Survey Database are displayed.

Checking Out Drawings and Data

When you check a file out of Vault to edit it, a copy of the file is placed in your working folder. The working folder is the local directory where the files on which you can work on are stored. The Check Out command gives you exclusive access to the data; no one else can check it out while you are working on it.

When you modify data, the project item state icon indicates the current status of the master file in the Vault. In the following illustration, the checkmark in the white circle for *maplewood_site.dwg* indicates that the drawing is checked out, and the local copy is the same as the version in the Vault. The checkmark in the green circle for parcel_boundary.dwg indicates that it is checked out to you, and the local copy is newer than the version in the Vault. This file should be checked back in to the Vault.

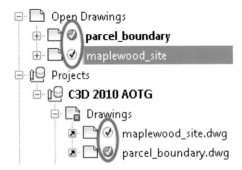

Check Out Command

To check a drawing out of Vault, in Toolspace, Prospector, right-click the drawing you want to check out. Select Check Out.

Check Out Drawing Dialog Box

The Check Out dialog box contains four sections, as shown in the following illustration.

① Specifies whether dependent files should be displayed.

② Specifies whether or not local copies of files are overwritten. If the option is selected, the latest version in the project database is checked out, overwriting the related file in the working folder.

③ Specifies the files and the dependent files, if applicable, that you want to check out.

④ Specifies an optional description of the file version you are checking in to the project. This will be the default comment when the file is checked in.

Procedure: Using the Check In Command

To check a drawing in to Vault, in Toolspace, Prospector, right-click the drawing. Click Check In.

Check In Drawing Dialog Box

The Check In Drawing dialog box contains four sections, as shown in the following illustration.

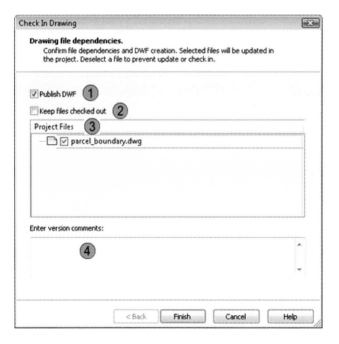

① Specifies drawings that can have DWF™ versions created or updated in the project.

② Specifies that the files are checked in and the versions are updated, but the drawing remains checked out to the working folder and available for revisions.

③ Specifies which files will be checked in to the project. If you do not want to check in a file, clear its check box.

④ Specifies an optional description of the files you are checking in to the project.

Exercise 01 | Work with Vault

In this exercise, you log in to Autodesk Vault from within Civil 3D. You also launch Vault Explorer, define a working folder, and make two new folders for the exercises in this lesson.

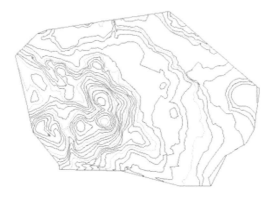

The completed exercise

Note: This exercise assumes that you have installed Autodesk Vault Server 2010, and that it is setup up with the following login entries:

- User Name: Administrator.

- Password: (blank).

- Server: localhost.

- Database: Vault.

1   Start AutoCAD Civil 3D.

2   In Toolspace:

   - Set Prospector to Master View.

   - Right-click Projects, click Log In to Vault.

3   In the Welcome dialog box, click Log In.

4   In the Log In dialog box:

- For User Name, enter **Administrator**.

- For Password, leave the field blank.

- For server, enter **localhost**.

- For Database, select Vault.

- Click OK.

5   In Prospector, right-click Projects. Click New.

6   Next, you log into the Vault Data Management Server, a separate application.

In the New Project dialog box:

- For Name, enter **C3D 2010 AOTG**.

- For Description, enter **AutoCAD Civil 3D 2010 Training Guidelines**, **Vault Lesson**.

- Click OK.

The new project is displayed in the Prospector window.

7   In Prospector, right-click Projects. Click Autodesk Vault.

8   In the Welcome dialog box, log in to Vault Explorer using the same details as in step 4.

9   Return to Civil 3D.

10   Open *Manage Data\maplewood_site.dwg*.

11   In Prospector, under Open Drawings:

- Right-click maplewood_site.

- Click Add to Project.

12   In the Add to Project wizard, Select a Project page, click Next.

13   On the Select a Drawing Location page, click Next.

14 On the Drawing File Dependencies page:

- Select the Publish DWF check box.

- Clear the Keep Files Checked Out check box.

- Click Next.

15 On the Share data page, check Surfaces. Click Finish.

The drawing is closed and added to the C3D 2010 AOTG Vault project.

16 Switch to Autodesk Vault 2010.

17 Return to Civil 3D.

18 In Prospector:

- Expand Projects, C3D 2010 AOTG, Drawings.

- Right-click *maplewood_site.dwg*.

- Click Check Out.

19 In the Check Out Drawing dialog box, click OK.

The drawing is checked out to you. A local working copy of the drawing is placed in the working folder *C:\Civil 3D Projects\AOTG*.

20 Open...*Manage Data\parcel_boundary.dwg*.

21 Repeat steps 10 through 19 to add drawing to the C3D 2010 AOTG project. This time, for the Drawing File Dependencies step, select the Keep Files Checked Out check box.

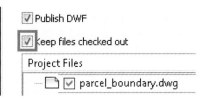

22 Make *maplewood_site* the current drawing.

Note: You can press CTRL-TAB on the keyboard to switch between drawings.

Next, you attach an external reference (Xref).

23 On the ribbon, Insert tab, Reference panel, click Attach.

24 In the Select Reference File dialog box:

- Browse to *Civil 3D Project\C3D 2010 AOTG*.

- Select *parcel_boundary.dwg*.

- Click Open.

25 In the Attach External Reference dialog box, for Insertion Point, clear the Specify on Screen checkbox. Click OK.

26 Make *parcel_boundary.dwg* the current drawing.

27 Modify the boundary as shown in the following illustration.

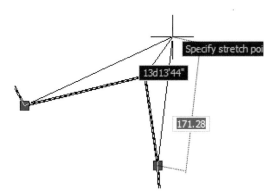

28 Save the file. The project and drawing item state icons change to indicate the status of the files.

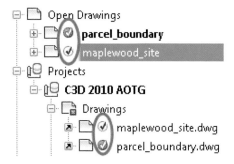

29 In Prospector, under Open Drawings, right-click *parcel_boundary*. Click Check In.

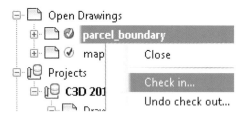

30 In the Check In Drawing dialog box, clear the Publish DWF and Keep Files Checked Out check boxes. Click Finish.

31 In the External References notification bubble, click Reload Parcel Boundary.

32 Save *maplewood_site*.

33 Check in *maplewood_site*. Finally, you reference a surface object from Vault.

34 Create a new drawing using the *_AutoCAD Civil 3D (Imperial) NCS.dwt*.

35 Save the drawing in the Manage Data folder as *Surfaces.dwg*.

36 In Prospector, expand Projects, C3D 2010 AOTG, Surfaces.

37 Right-click Phase 6 Orig Ground. Click Create Reference.

38 In the Create Surface Reference dialog box, click OK.

39 Zoom in to the extents of the drawing and note the surface.

The surface was referenced from Vault. If the surface data changes in the original drawing, the surface automatically updates in this drawing.

The completed exercise drawing appears as follows:

40 Close all files. Do not save. Exit Civil 3D.

Index

A

Active data shortcuts
  project 652
Adjoining parcels 220
Alignment command
  settings 270
Alignment geometry 261
Alignment objects 261
Alignment stationing 315
Alignment superelevation 528
Alignment table 269
Alignment Tag Labels 269
Alignment, entities 485
Alignments and Parcels 211
Alignments from Objects,
  Creating 260
Alignments, definition 261
Alignments, editing 486
Alignments, editing
  in tables 488
Alternate Point
  Descriptions 150
Ambient Settings 49
Analyze a surface 193
Assemblies 327
Assemblies, Creating 331
Assembly Elements 328
Assembly groups 332
Assembly insertion
  frequency 348
Assembly object 326
Assembly Properties,
  Modifying 535
Assembly set 327, 576
Associated annotation 28, 304
AutoCAD Elements 48
AutoCAD Settings 49
Autodesk Vault 678

Autodesk Vault 2010 Data
  Management Server 678
Automatic recalculation 119
Average Breakline
  Elevation 180

B

Base stationing values 262
Baseline 328, 344
Baseline point 328
BasicSideSlopeCutDitch 330
Begin command 115
Bisect 220
Boundary analysis 86
Boundary Options 564
Boundary survey 86, 211
Breakline 169, 171

C

Calculating Corridor
  Quantities 610, 613
Calculating Earthwork
  Volumes 401
Calculating Quantity Takeoff
  Using Pay Items 664
Centerline alignment 512, 515
Central repository 76, 679
Check Out command 686
Check Out Drawing
  Dialog Box 687
Checking Out Drawings
  and Data 686
Circular grips 306
Civil 3D Elements 48
Civil 3D surface object 361
Client-side Components 680

Code Set Styles 589, 668
Command line 5
Command settings 28, 36, 50
Commands 3
Comparison surface 399
Composite surface 401
Construction sheets 639
Continuous corridor 345
Continuous surface 167
Control points 77, 87
Coordinate zone 19
Corridor 32
Corridor Boundaries 564
Corridor configuration 343
Corridor data 362
Corridor datum 610
Corridor extents 564
Corridor feature lines 369
Corridor model 326, 347
Corridor Models, Creating 342
Corridor Models,
  Definition 343
Corridor properties 362
Corridor quantities 610
Corridor regions 348
Corridor surfaces 361
Corridor Surfaces,
  Creating 360
Corridor Surfaces,
  Creating 360
Corridor, Complex 345
Corridors collection 347
Create Assembly command 331
Create Intersection
  wizard 570, 573
Create ROW 211
Create Sheets Wizard 639
Create Surface dialog 400

Create View Frames
 Wizard 639
Creating Assemblies 326
Creating intersections,
 process 577
Creating Quantity
 Reports 620
Crest curves 315
Criteria-based alignments,
 designing 484
Criteria-Based Design
 489, 490
Cross section-based
 features 326
Cross-section elements 327
Crossing Breaklines 180
Cul-de-sac 345, 348
Curb return parameters 571
Custom subassembly 327
Cyan triangle grips 306

D

Data collector files 77
Data Points 281
Data Shortcuts 650
Data Shortcuts manager 651
Data Shortcuts, Definition 651
Datum links 363
Daylight feature line 546
Daylight line 529
Daylighting 529
Daylighting subassemblies
 329, 528
Definition of Section
 Views 629
Deleting Triangles 181
Depth labels 314

Description Keys 146, 149
Design Checks 490
Design Criteria File 490
Design layout profiles 290
Design objects 203
Design speed 505
Dimension Anchor 317
Direction lines 87
Display a surface 193
Display parameters 316
Document management
 system 679
Draw Parts in Profile 449
Drawing Settings 49
Drawing template 149, 327
Drawing Templates,
 Creating 46
Drawing Templates,
 Definition 47
Drawing templates,
 preconfigured 47
Driving Direction 571
Dynamic alignment 513
Dynamic offset alignment 513
Dynamic profiles 281
Dynamic relationships 30
Dynamic surface labels 415

E

Edit Profile Geometry 308
Edit survey observations 20
Editing with Layout Tools 308
EG FG Combined surface 591
Error ellipses 87
Existing Ground surface 399
Expansion factor 612
Export points 147

External reports 17
Extract section data 599

F

Feature command 50
Feature line points 369
Feature line segments 369
Feature Lines 344
Feature Lines from
 Surfaces 384
Feature Lines, Creating 368
Feature Lines, Definition 369
Feature settings 28
Field book file 78
Figure groups 115
Figure prefix database 95, 98
Figure prefixes 95
Figure Styles and Prefixes,
 Creating 94
Figures, definition 115
Fill factors 611
Fill volume 563
Final Grading Surface,
 Labeling 414
Final Grading Surfaces,
 Creating 396
Finished grade surface 563
Flip Label 250
Frontage offset 221

G

Geometry data 248
GPS data files 87, 105
Grade breaklines 315
Grade to Surface 399
Grading criteria 398

Grading Feature Lines 370, 382
Grading groups 397, 398
Grading Object Style 35
Grading Objects 397
Grading Objects, Creating 398
Graphical Editing 305
Graphical editing grips 305
Grid objects 280
Grip edit 239
Grip Types 137
Grips 136

H

Horizontal alignment 261, 280
Horizontal axes 316
Hydraflow Hydrographs
  Extension 471
Hydraflow Storm Sewers
  Extension 468

I

Import survey data 20, 104
Importing and Creating
  Points 132
Importing the Field
  Book File 106
Input parameters 28, 31
Insert PVIs Tabular 292
Instrument setups 87
Interim grading surface 343
Interim Grading Surfaces,
  Creating 380
Intersections, Creating 570
Intersections, Definition 571

K

K values 304

L

Label Sets 316
Label styles 17
Label Styles, Creating 28
Label Styles, Definition 33
Label Styles, Types 34
Labeling Parcel Segments 248
Labeling Profiles 314
Land survey 76
Land XML file 119
Lane subassemblies 329
Lane width target 330
LaneOutsideSuper 329
Layout profile data 308
Layout profile geometry 291
Layout Profiles 290
Legal Survey and Survey
  Control 134
Limit of construction 529
Linework connectivity 109
Linework connectivity
  codes 95
Local survey database 76

M

Major Contour 194
Managing Drawings 685
Managing Points 146
Manning n value 474
Mass haul lines 599
Master copy 682
Maximum Triangle Length 178
Measured horizontal angles 115
Minor Contour 194
Modeling Road Designs
  in 3D 588
Multiple section views 630

N

National CAD Standard 47
Network Layout Tools 431
Network lines 87
Network parts 428
Null Structures 433

O

Object label styles 19
Object naming
  templates 28, 31
Object reactivity 30
Object styles 19
Object Styles, Creating 28, 37
Object Styles, Definition 33
Object-oriented design 29
Objects Used to
  Create Points 139
Objects, Creating 28
Observation Data 116
Observation measurements 77
Offset Alignments 513
Offset Alignments, Create 512
Offset feature lines 382
Open FIles 9
Operation Type List 178
Organize Data 204

P

Panels 3
Panorama 304
Panorama window 308
Parcel area labels 219
Parcel area report 64
Parcel Creation Tools 219
Parcel editing commands 240
Parcel geometry 239
Parcel Geometry, Editing 239
Parcel Inverse report 62
Parcel Layout tools 218, 225

Parcel Layout Tools 220, 239
Parcel numbering 239
Parcel objects 219
Parcel parameters, modify 212
Parcel segment labels 36, 248
Parcel segments 219
Parcel Style 35
Parcel tables 251
Parcel tags 251
Parcels, Creating 218
Parcels, Editing 238
Parent alignment 515
Parent horizontal
  alignment 315
Part Creation Modes 432
Part Rules 431
Parts List 431
Pavement structure 599
Pavement structure
  quantities 610
Pay item codes 666
Pay item file 666
Pay item list 667
Pay items 664
Pay Items, Assigning 667
Perspective views 590
Photorealistic materials 588
Pipe and Structure Rules 431
Pipe drop 432
Pipe Labels 461
Pipe Network Catalog 431
Pipe Network Creation
  Tools 431
Pipe Network
  Relationships 428
Pipe Networks in
  Profile View 449
Pipe Networks,
  Creating 427, 435
Pipe Networks, Drawing 448
Pipe Networks, Editing 448

Pipe Properties 451
Pipe slopes 449
Pipe styles 452
Pipe, Data Table Edits 450
Pipes, Labeling 460
Plan Production 639
Point data, text file 134
Point descriptions 109
Point Groups, Definition 147
Point label style 147, 150
Point style 147, 150
Point table style 148
Point tables 146, 148
Points and Figures,
  Creating 117
Points Created with
  a Parcel 140
Points Created with
  an Alignment 141
Points, Definition 133
Predefined subassemblies 327
Profile 32
Profile Editor 282
Profile Geometry 304
Profile Geometry, Editing 305
Profile Labels, Editing 317
Profile Layout Tools 308
Profile object 305
Profile View Labels 316
Profile View Style 283, 314, 316
Profile view title 314
Profile Views 280, 282, 314
Project data 17
Project members 682
Project object collections 682
Prospector 18
Prospector tab 18
Proximity Breaklines 169
PVI 291
PVI elevations 292
PVI stations 292

Q

Quantity Report Types 621
Quantity Takeoff 665, 668
Quantity Takeoff Criteria 611

R

Random points 139
Raw survey data 107, 115
Red triangle grip 306
Reference Objects 650
Reference Objects,
  Definition 651
Refill factors 611
Regions 344
Regular surfaces 167
Render material 589
Renumbering Parcels 240
Report settings 65
Reports Manager 63
Reports, Creating 62
Reports, Format 66
Resize parcels 220
Reverse Label 250
Ribbon 3
Ribbons, Contextual 11
Ribbons, Static 10
Right-of-way parcel 211
Right-of-Way Parcels,
  Creating 210
Road centerline
  alignments 261
Road profiles 281

S

Sag curves 315
Sample line 600
Sample Line Groups 599
Sample Line Parameters 600
Sample Lines, Creating 598
Sample Lines, Moving 602
Scale 19
Section view 630
Section View, Groups 629
Section View, Style 629
Section views 631
Section Views, Creating 628
Segment geometry 248
Server-side Component 680
Set Path 98
Settings tab 19, 24
Sheets, Creating 641
ShortCutNode command 653
Sideshot lines 87
Sideshot points 87
Site analysis 86
Sites 203
Sites, Creating 202
Sites, Definition 203
Slope assembly 529
Slope distances 115
Slope parameter 529
Spanning labels 462
Spatial relationship 204
Spot Elevation Surface
  Labels 416
Spot elevations 139
Square grip 306
Stakeout information,
  create 140
Standard Breaklines 169
Standard curve settings 292
Standard deviations 77
Starting station 262
Station and curve report 64
Station reference point 269

Stepped Offset 369, 370
Storm Sewer Networks,
  Designing 468, 472
Storm sewer, design
  calculations 469
Storm sewer, pipe data 471
Structure label styles 461
Style assignment 28
Subassemblies 327
Subassembly catalog 327, 331
Subassembly Components 531
Subassembly Groups 534
Subassembly Links 533
Subassembly markers 332
Subassembly Parameters 328
Subassembly Points 532
Subassembly Shapes 533
Subassembly,
  BasicGuardrail 532
Subassembly.
  LaneOutsideSuper 532
Subdivide 220
Subdivision Corridor
  Models 343
Subgrade surfaces 563
Summary Takeoff report 665
Superelevation properties 504
Superelevation region 504
Superelevation tables 504
Superelevation, applying 504
Surface Annotation 415
Surface Boundary 361
Surface Display Styles 193
Surface labels 414
Surface Profiles 280
Surface Properties 177
Surface Properties dialog 180
Surface style 193
Surface Style dialog 194
Surface Styles, Creating 192
Surface Triangles 177

Surface Triangles, Editing 181
Surfaces from point data 171
Surfaces, Creating 166
Surfaces, Modifying 176
Survey command window 77
Survey Data
  Characteristics 116
Survey Data Collection
  Link 107
Survey Data, Working with 114
Survey database 77, 79, 87
Survey Database Settings 80
Survey databases 20
Survey Databases, Creating 76
Survey editors 77
Survey field book 105
Survey figure 96
Survey Network 31, 88
Survey network style 87
Survey Network,
  recalculate 116
Survey networks 20
Survey Networks
  Components 89
Survey Networks, Creating 86
Survey observation 20
Survey observation data 77, 115
Survey observations 87
Survey tab 20
Survey User Settings 81, 82
Swapping Triangle Edges 182
Synchronize reference
  objects 651

T

Tabs 3
Tag label style 269
Tag labels 252, 270
Tags 248
Tangent grades 292
Tangents 290
Target alignment 547
Target Mapping 546
Target Mapping Lane
  Widths 547
Target surface 399
Terrain data 281
TIN lines 167
TIN Volume Surfaces 400
TIN Volume Surfaces,
  Definition 400
Toolbars 5
Toolbox tab 20
Toolspace 16
Toolspace Tabs 18
Top Corridor Surface 362
Topographic survey 86, 134
Topology 203, 204
Total station 87, 105
Transformation Settings 49
Transparent commands
  293, 294
Transparent commands 309
Transportation assemblies,
  modifying 528
Transportation corridor
  models 529
Transportation Corridor
  Models, Creating 548
Transportation Corridor
  Surfaces, Creating 562
Transportation Corridors,
  Creating 544
Transportation design 485
Transportation palette 531

Traverse definitions 77
Triangulated Irregular
  Network 167
Types of external 63

U

Units 19
User interface 2

V

Vault, Accessing 682
Vertical angles 115
Vertical axes 316
Vertical curve lengths 292
Vertical curves 290
Vertical exaggeration 284
Vertical geometry 304
View frame group 641
View Frames, Creating 640
Visualize a surface 193
Volume data 398
Volume surfaces 167

W

Warning marker 491
Widening parameters 513
Widenings 513
Working folder 686
Workspace, Create 14

Notes

Notes